VOLUME ONE HUNDRED AND TWENTY THREE

Advances in IMMUNOLOGY

Cell Biology of the B Cell Receptor

ASSOCIATE EDITORS

VOLUME ONE HUNDRED AND TWENTY THREE

ADVANCES IN IMMUNOLOGY

Cell Biology of the B Cell Receptor

Edited by

HIDDE L. PLOEGH

Whitehead Institute for Biomedical Research,
Cambridge,
Massachusetts, USA

Series Editor

FREDERICK W. ALT

Howard Hughes Medical Institute,
Boston,
Massachusetts, USA

ELSEVIER

AMSTERDAM • BOSTON • HEIDELBERG • LONDON
NEW YORK • OXFORD • PARIS • SAN DIEGO
SAN FRANCISCO • SINGAPORE • SYDNEY • TOKYO
Academic Press is an imprint of Elsevier

AP

Academic Press is an imprint of Elsevier
525 B Street, Suite 1800, San Diego, CA 92101-4495, USA
225 Wyman Street, Waltham, MA 02451, USA
The Boulevard, Langford Lane, Kidlington, Oxford, OX5 1GB, UK
32 Jamestown Road, London, NW1 7BY, UK
Radarweg 29, PO Box 211, 1000 AE Amsterdam, The Netherlands

First edition 2014

ISBN: 978-0-12-800266-7
ISSN: 0065-2776

For information on all Academic Press publications
visit our website at store.elsevier.com

CONTENTS

CONTRIBUTORS

Ana M. Avalos
Whitehead Institute for Biomedical Research, Cambridge, Massachusetts, USA

Hassan Jumaa
Spemann Graduate School of Biology and Medicine (SGBM), Albert Ludwigs University Freiburg; Department of Molecular immunology, Faculty of Biology, University of Freiburg and Max Planck Institute of Immunobiology and Epigenetics, Freiburg, and Department of Immunology, Ulm University, Ulm, Germany

Ana-Maria Lennon-Duménil
INSERM U932, Institut Curie, Paris, France

Friederike Meyer-Wentrup
Department of Pediatric Hematology and Oncology, Wilhelmina Children's Hospital, University Medical Center Utrecht, Utrecht, The Netherlands

Dorian Obino
INSERM U932, Institut Curie, Paris, France

Hidde L. Ploegh
Whitehead Institute for Biomedical Research, Cambridge, Massachusetts, USA

Katelyn M. Spillane
Division of Immune Cell Biology, National Institute for Medical Research, London, United Kingdom

Elena Surova
Spemann Graduate School of Biology and Medicine (SGBM), Albert Ludwigs University Freiburg, and Department of Molecular immunology, Faculty of Biology, University of Freiburg and Max Planck Institute of Immunobiology and Epigenetics, Freiburg, Germany

Pavel Tolar
Division of Immune Cell Biology, National Institute for Medical Research, London, United Kingdom

PREFACE

The interest in various immunological subdisciplines waxes and wanes with every new discovery. On the antibody front, much has been written recently about broadly neutralizing antibodies against viruses such as HIV and influenza. This has spawned immense interest in somatic hypermutation, follicular helper T cells, and the germinal center reaction as a gateway to the design of improved vaccines, especially in cases where antigenic variation in the pathogen seemingly outpaces the ability of the adaptive immune response to deploy effective countermeasures. Nonetheless, before such processes can kick in and all of the relevant cell types have been called into action, many steps are required to ensure proper initial engagement of antigen-specific B cells. This volume deals with some of these upstream events, mostly but not exclusively from a cell biological perspective. The chapter by Avalos et al., the first in this volume, sets the stage by painting current knowledge of B cell activation with a rather broad brush and on this canvas presents some of the more recent observations that derive from new mouse models, created by somatic cell nuclear transfer using nuclei from antigen-specific B cells as donors. The chapter further seeks to draw parallels between normal B cell activation and the events that are believed to predispose to malignant transformation. This topic is picked up in the chapter by Jumaa, whose discoveries on the spontaneous, seemingly antigen-independent activation of B cells may help explain at least some of the predispositions to malignant transformation. A picture often convinces more readily than the text required to explain it. Especially in the field of lymphocyte activation, high-end imaging techniques continue to shape our thinking about the interactions that drive a properly orchestrated immune response. The chapter by Tolar draws on his extensive experience in imaging early events in the engagement of the B cell receptor. A careful time-resolved analysis of early B cell activation events, with emphasis on the BCR, provides a vista that is not readily accessible by more conventional methods such as cytofluorimetry. Especially the spatial integration of different signaling pathways must be addressed by morphological analysis as well as biochemistry, if we are ever to make sense of the multiple signals that impinge on B cells and their proper interpretation. Finally, moving from the BCR to some of its downstream effectors, the role of the cytoskeleton in organizing a platform that renders a B cell competent to signal has become part of immunological canon. The chapter by Lennon-Duménil reviews recent advances in this area and emphasizes the notion that we stand to benefit by taking a more integrated view of the B cell, ensuring proper connections between the first encounter with antigen, the role of an organized, membrane-bound platform to process such signals and transmit them to downstream effectors. Combined, the chapters in this volume cover this important topic by reviewing the most recent and relevant literature on the biology of the B cell receptor.

Hidde L. Ploegh

CHAPTER ONE

B-Cell Receptor Signaling in Lymphoid Malignancies and Autoimmunity

Ana M. Avalos*, Friederike Meyer-Wentrup†, Hidde L. Ploegh*,[1]
*Whitehead Institute for Biomedical Research, Cambridge, Massachusetts, USA
†Department of Pediatric Hematology and Oncology, Wilhelmina Children's Hospital, University Medical Center Utrecht, Utrecht, The Netherlands
[1]Corresponding author: e-mail address: ploegh@wi.mit.edu

Contents

Advances in Immunology, Volume 123
ISSN 0065-2776
http://dx.doi.org/10.1016/B978-0-12-800266-7.00004-2

Abstract

The B-cell receptor (BCR) for antigen is a key sensor required for B-cell development, survival, and activation. Rigorous selection checkpoints ensure that the mature B-cell compartment in the periphery is largely purged of self-reactive B cells. However, autoreactive B cells escape selection and persist in the periphery as anergic or clonally ignorant B cells. Under the influence of genetic or environmental factors, which are not completely understood, autoreactive B cells may be activated. Similar activation can also occur at different stages of B-cell maturation in the bone marrow or in peripheral lymphoid organs and give rise to malignant B cells. The pathology that typifies neoplastic lymphocytes and autoreactive B cells differs: malignant B cells proliferate and occupy niches otherwise taken up by healthy leukocytes or erythrocytes, while autoreactive B cells produce pathogenic antibodies or present self-antigen to T cells. However, both malignant and autoreactive B cells share the commonality of deregulated BCR pathways as principal contributors to pathogenicity. We first summarize current views of BCR activation. We then explore how anomalous BCR pathways correlate with malignancies and autoimmunity. We also elaborate on the activation of TLR pathways in abnormal B cells and how they contribute to maintenance of pathology. Finally, we outline the benefits and emergence of mouse models generated by somatic cell nuclear transfer to study B-cell function in manners for which current transgenic models may be less well suited.

1. INTRODUCTION: B-CELL BEHAVIOR IN AUTOIMMUNITY AND B-CELL MALIGNANCIES

In both autoimmune disease and B-cell malignancies, B cells exhibit deregulated behavior. The B-cell receptor (BCR) for antigen is a sensor that must discriminate signals perceived through the antigen-binding site to determine whether a B cell bearing such a BCR should survive, enter the cell cycle, or die by apoptosis. Not surprisingly, altered BCR signaling can be found in autoreactive and malignant B cells. In this chapter, we outline the main alterations in BCR signaling pathways linked to autoimmunity and B-cell lymphomas and leukemia. Admittedly, it is not possible to cover all possible aspects within the context of this chapter. We summarize processes that could go awry in B-cell development and activation—events intimately linked to the BCR—that can lead to these disorders. We examine the TLR pathways that may be engaged in these diseases to enhance the proliferative and activation capacity of deregulated B cells.

Much has been learned in the recent past about the cell biology of BCRs, with significant new results pertaining to the role of their interactions with cytoskeletal elements and their movement in response to ligation with antigen from detailed intracellular analyses using superresolution microscopy and sophisticated biochemical methods. The ability of a BCR to display properties of a fully activated receptor, even in the absence of a detectable cognate antigen, may be an important driver for malignant transformation, as such antigen-independent signals would be perceived persistently and might almost justify the qualification of such BCRs as oncogenes. The three reviews that follow in this volume deal with these aspects. The present chapter is intended to provide a broader conceptual framework that will help place these chapters in perspective.

We first define a prototype systemic autoimmune disease, systemic lupus erythematosus (SLE), and known B-cell malignancies. We then describe the main B-cell developmental steps and tolerance checkpoints that prevent autoreactive specificities in the periphery. We then explore the initial cues required for B-cell activation in a normal setting, as well as the BCR pathways activated by antigen. We explore the BCR pathways that can be mutated or altered in SLE and B-cell malignancies, as well as the role of TLR signaling in aiding breakdown of tolerance in these settings. We outline how self-antigen and tonic signaling support persistence of neoplastic B cells. Finally, we compare SLE and B-cell malignancies with regard to the approach used to study them and their underlying pathology.

To expand the toolkit used to study B-cell activation in general, we summarize results obtained with new mouse models, created by somatic cell nuclear transfer (SCNT), which demonstrate that isotypes other than IgM fully support B-cell development. These models also contribute to our knowledge of the valency requirements of B-cell recognition of antigen and can address in ways not previously possible secondary class-switch recombination (CSR). SCNT allows the production of mice homogeneous with respect to their BCRs (or TCRs), starting from very few antigen-specific lymphocytes. In principle, this opens up opportunities to study rare autoimmune diseases by using as nucleus donors tissue-resident cells that contribute to pathology, as well as discerning novel antigens that may drive neoplastic B-cell expansion. The emerging use of SCNT to examine aspects of B-cell biology not previously accessible will expand our grasp on how deregulation in autoimmunity and B-cell neoplasias occurs.

2. SYSTEMIC LUPUS ERYTHEMATOSUS

To study the role of BCR pathways in autoimmunity, we selected a prototype systemic autoimmune disease that has been extensively studied and whose pathology is B cell-dependent. SLE is a polygenic autoimmune disease characterized by polyclonal B-cell activation, hypergammaglobulinemia with circulating antibodies to DNA and ribonucleoprotein (RNP), immune complex deposition in kidneys, and, eventually, end organ damage (Rahman & Isenberg, 2008). The pathogenesis of SLE manifests itself in several systemic symptoms, which range from skin involvement, to joint inflammation, to organ involvement, typically due to the formation of immune complexes with circulating autoantigens and deposition in the kidneys, small vessels, and, in some cases, the brain. The causes of SLE are unknown, yet environmental, hormonal, genetic, and immunologic triggers may act in conjunction to promote disease. Environmental factors can include UV radiation, drugs that trigger DNA hypomethylation, and viruses; a link between Epstein–Barr virus (EBV) infection and lupus onset has been reported (Rahman & Isenberg, 2008). The high prevalence of lupus in females/males (9:1) suggests that hormonal factors may also play a role in disease onset (Grimaldi, Hicks, & Diamond, 2005). Current treatments for SLE do not provide permanent remedy but alleviate symptoms and are tailored to the patients' main signs and include anti-inflammatory drugs and antimalarials. The major causes of death in patients with SLE are complications of treatment, such as infections and atherosclerosis (Ward, Pyun, & Studenski, 1995). In recent years, the use of biologics that aim to modulate key effectors of dysregulated B-cell biology is gaining strength, with reagents directed to B cells currently in clinical trials. Such therapeutics include anti-CD20 (rituximab), a monoclonal antibody directed to CD20, a receptor involved in B-cell cycle expressed on mature B cells; anti-CD22 (epratuzumab), a monoclonal antibody specific for the CD22 inhibitory receptor that modulates BCR responses; and inhibitors of B cell-activating factor (BAFF) (belimumab), a soluble B-cell survival factor whose levels are upregulated in patients with SLE (Paz & Tsokos, 2013).

3. B-CELL MALIGNANCIES

B-cell malignancies manifest themselves as leukemia or lymphoma, depending on the site of primary presentation (bone marrow in case of leukemia and a mass in case of lymphoma). The World Health Organization (WHO) classification stratifies B-cell malignancies according to the characteristics of the cell of origin such as immune phenotype, cytogenetics, and morphology (Swerdlow et al., 2008). Clinically, B-cell malignancies can be classified according to their proliferative potential and biological behavior as indolent, aggressive, and highly aggressive. Hodgkin lymphoma is a B lymphoid malignancy, the clinical presentation, biology, and prognosis of which are different from other B-cell malignancies, and is therefore considered a distinct entity. B-cell lymphomas are thus grouped into Hodgkin and non-Hodgkin lymphomas (HL and NHL). The incidence of the different types of B-cell malignancies differs significantly between adults and children. The latter usually present with acute precursor B-cell leukemia (B-ALL), Burkitt lymphoma, and diffuse large B-cell lymphoma (DLBCL). Precursor B-cell leukemia and Burkitt lymphoma show highly aggressive behavior and DLBCL shows aggressive clinical behavior. In adults, the most common B-cell neoplasms are chronic lymphatic B-cell leukemia (CLL), follicular lymphoma, and DLBCL. CLL derives from naive B cells, while DLBCL and follicular lymphoma originate from germinal center (GC) B cells. Their clinical behavior is generally less aggressive than that of B-cell neoplasms found in children. Even though the diagnosis of DLBCL in children and adults is made according to the same WHO classification criteria, DLBCLs in adults and children differ in their biological properties (Klapper et al., 2008; Miles et al., 2008). Pediatric DLBCL responds well to therapy and has an excellent prognosis (Miles et al., 2008; Oschlies et al., 2006). In adults, DLBCL is a heterogeneous disease with a far poorer clinical outcome (Rosenwald et al., 2002). The differences might be explained by the predominance of the germinal center B-cell (GCB) subtype in children (83%) and in adults (50%), which in adults is a favorable prognostic factor. Expression of surface markers by the GCB subtype differs significantly between children and adults. While 70% of pediatric cases are CD10-positive, this is the case in only 30% of adult GC-type DLBCL. BCL6 positivity is detected in 90% of pediatric cases versus 70% of adult cases (Oschlies et al., 2006). BCL6 alters expression of genes characteristically expressed by GC B cells: a blockade of BCL6 function results in downregulation of CD10 and upregulation of CD38, a marker of terminal plasmacytic differentiation in the GC (Shaffer et al., 2000).

Burkitt lymphomas in children and adults comprise one disease entity with similar clinical behavior and immune phenotype (Miles et al., 2008). They can occur sporadically or related to infection with the EBV. Almost all B-cell neoplasias, except for classical Hodgkin lymphoma and the very immature pediatric B-ALLs, express cell surface BCR. In pediatric B-ALL, tonic pre-BCR signaling is essential for leukemogenesis. BCR signaling supports proliferation of neoplastic B cells on different cellular levels ranging from expression of stereotypic antigen-binding sites for recognition of exogenous and endogenous antigens to changes in the BCR signal transduction pathway. These findings have enabled the development of better-targeted therapies for B-cell leukemia and lymphoma. Current standard therapies rely on chemotherapeutic agents, surgery, and radiotherapy depending on patient age and type of leukemia/lymphoma. For adult patients, anti-CD20 antibody therapy has become the standard of care (rituximab) (Maloney, 2012). Selective kinase inhibitors are currently tested in phase I and II trials (see the succeeding text).

4. ROLE OF BCR IN B-CELL DEVELOPMENT AND SELECTION

The formation of a BCR by rearrangements of the *V, (D), J* gene segments in the heavy- and light-chain loci is the hallmark of B-cell development. Since such random gene joining events can lead to self-reactivity, selection steps reduce the number of self-reactive B cells in the repertoire. Because these checkpoints depend on interactions of the BCR with "self," the sensitivity of these interactions needs to be tightly regulated to distinguish signals for positive versus negative selection.

4.1. Stages of B-cell development

B cells develop from pluripotent hematopoietic stem cells in the fetal liver and in the bone marrow after birth (Hardy & Hayakawa, 2001). B cells develop throughout life and undergo different stages of differentiation until they become mature B cells ready to encounter antigens in the periphery. The rearrangement of the immunoglobulin (Ig) heavy- and light-chain loci defines the stages of B-cell development (Fig. 1.1), further characterized by the expression of diagnostic surface markers. A common lymphoid progenitor gives rise to the B lineage-committed precursor, which lacks immunoglobulin gene rearrangements and is referred to as a pro-B cell. Rearrangements on the *V, D,* and *J* genes of the heavy-chain locus (to give rise to variable regions of Ig) occur upon introduction of double-stranded breaks by recombination-activating gene (RAG) proteins 1 and 2 and ancillary proteins involved in somatic gene rearrangement. Splicing of the primary transcript joins the *VDJ* element to its downstream constant region, typically μ. The newly formed H chain then pairs with the surrogate light chains, λ5 and VpreB, to yield the signaling-competent pre-B-cell receptor at the pre-B-cell stage (Fig. 1.1). The pre-B-cell receptor associates with Igα and Igβ, which contain the immunoreceptor tyrosine-based activation motifs (ITAMs) required for transduction of signals (Hardy & Hayakawa, 2001). Assembly of pre-BCR is accompanied by a burst of proliferation, which is promoted by interactions of the pre-BCR with self-molecules in stromal cells. Such proliferation of pre-B cells bearing self-reactive BCRs serves to expand the pool of B cells

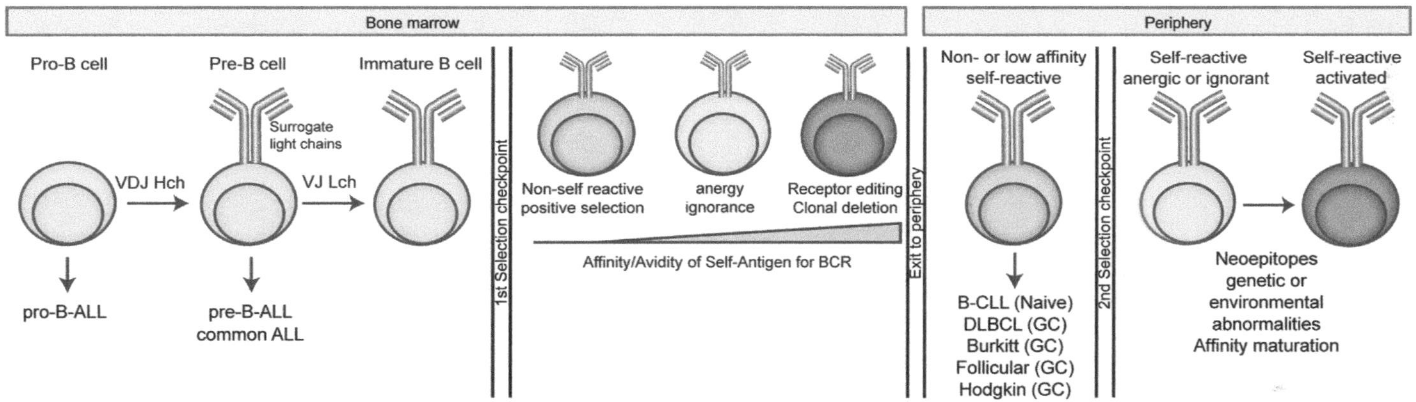

Figure 1.1 *Different stages of normal B-cell development can give rise to autoreactive and malignant B cells.* Immature pre- and pro-B cells of the bone marrow can give rise to pro- and pre-B-cell acute lymphoblastic leukemia (B-ALL), as well as common ALL. Depending on the affinity/avidity of self-antigen for immature B cell, autoreactive B cells can be deleted or tolerized. Upon exit to the periphery, residual autoreactive specificities can undergo a second round of selection, and autoreactive B cells can be tolerized or deleted. Circulating low-affinity self-reactive B cells can be activated by encountering new epitopes or by environmental or genetic predisposition. Naive B cells can give rise to lymphatic B-cell leukemia (B-CLL), while autoreactive and malignant B cells can arise after a germinal center (GC) reaction promoting diffuse large B-cell lymphoma (DLBCL) and Burkitt, follicular, or Hodgkin lymphoma. (See the color plate.)

containing rearranged heavy-chain loci that will undergo light-chain rearrangements, thus enlarging the immature B-cell repertoire (Herzog & Jumaa, 2012). Proliferation is accompanied by a transient downregulation of RAG1/2, before reactivation of the recombination machinery to enable *V* to *J* rearrangement at the light-chain locus to produce a mature BCR at the immature B-cell stage. The gene rearrangements that occur in the course of B-cell development constitute a genetic vulnerability and can produce translocations with concomitant activation of oncogenes and cause malignancies (Tsujimoto, Cossman, Jaffe, & Croce, 1985), as will be discussed in the succeeding text. B-cell leukemia in infants and children is usually derived from the very early B-cell progenitors, the pre- and pro-B cells (Fig. 1.1).

4.2. Central tolerance mechanisms

After formation of a BCR on the surface of immature B cells, self-reactive B cells are negatively selected based on their interaction with self-molecules in the bone marrow. From an initial 75% of immature B cells in the bone marrow that display some measure of specificity for self, about 50% of newly emergent B cells show autoreactivity, and circulating B cells show about 20% self-reactivity (Wardemann et al., 2003). Thus, two checkpoints in the bone marrow and in the periphery ensure that autoreactive specificities are largely purged from the circulating B-cell repertoire (Fig. 1.1). Patients with autoimmune disorders such as SLE show defective tolerance at both stages (Meffre & Wardemann, 2008).

Four main central tolerance mechanisms have been identified, and B-cell clones that are persistently autoreactive are dealt with according to the strength of interaction with self: high-avidity/high-affinity interactions lead to receptor editing or clonal deletion, while low-affinity interactions induce anergy or ignorance. In the context of antigen recognition, avidity is defined as the overall strength for the antibody–antigen interaction, and it depends on both the affinity of the antibody for the antigen epitope (which is defined by the strength of interaction of antigen and antibody in equilibrium or $K_D = [Ab^*Ag]/[Ab][Ag]$) and the valency of interaction. A more detailed view of the valency requirements for activation is given later in this chapter. Even though BCR autoreactivity in B-cell malignancy has clearly been demonstrated (see in the succeeding text), the exact mechanisms of how tolerance is broken remain to be elucidated.

A relatively high percentage of high-avidity self-reactive B cells are generated during B-cell development, which die by apoptosis or undergo receptor editing. Receptor editing is attempted first, and it comprises secondary rearrangements of the light-chain locus to obtain non-self-reactive species: if interactions of BCR with self exceed a certain threshold, then secondary rearrangements of the light-chain locus occur to generate a heavy-/light-chain combination that is no longer autoreactive (Gay, Saunders, Camper, & Weigert, 1993; Radic, Erikson, Litwin, & Weigert, 1993; Tiegs, Russell, & Nemazee, 1993). If the process of receptor editing fails to render non-self-reactive BCRs, then the B cells are deleted from the repertoire by induction of apoptosis. This phenomenon was first shown in the anti-MHC class I transgenic B-cell model (Nemazee & Burki, 1989) and later in the hen egg lysozyme (HEL) transgenic BCR model, where the rearranged heavy- and light-chain loci of HEL-specific Ig were expressed together with a modified membrane-bound HEL construct. Extensive crosslinking of the HEL BCR by membrane-bound HEL leads to deletion of HEL-specific B cells (Hartley et al., 1991). These early studies introduced the concept that avidity of antigen for BCR determines the fate of autoreactive B cells.

In double-transgenic mice where soluble HEL was coexpressed with a HEL-specific BCR, HEL-specific B cells are present in the periphery, but their ability to secrete antibody was impaired, consistent with induction of anergy (Goodnow et al., 1988). Thus, in the HEL/anti-HEL model, B cells become anergic due to the lower avidity of interactions with soluble HEL (possibly recognized in monomeric form), while the increase in avidity afforded by membrane display of HEL led to clonal deletion. Systemic autoimmune diseases such as SLE are polygenic, yet they display a limited set of autoantigens, such as double-stranded and single-stranded DNAs, Smith (Sm) antigen, RNP, and gamma immunoglobulins (IgG) specific for these antigens (Shlomchik, 2008). Transgenic mice specific for DNA, Sm, and Ars1 (Ars1-specific BCR cross-reacts with DNA) display antigen-specific B cells in their periphery that are anergic (Erikson et al., 1991; Merrell et al., 2006; Santulli-Marotto, Retter, Gee, Mamula, & Clarke, 1998). Anergic cells have much reduced responses to antigen, for example, as assessed by calcium mobilization, and reduced IgM expression on the surface. Baseline calcium levels are elevated in anergic cells, suggesting that at least some kind of signal is being perceived (reviewed in Cambier, Gauld, Merrell, and Vilen (2007)).

A fourth type of tolerance induction is characterized by the presence of self-antigen-specific B cells in the periphery with BCRs of low affinity (and therefore, the cells that bear these BCRs escape receptor editing and clonal deletion) that otherwise behave as normal follicular B cells. This tolerance mechanism, termed clonal ignorance, was observed in a transgenic mouse specific for rheumatoid factor (IgG RF) that was allotype-restricted: the AM14 BCR is specific for IgG2a of the "a" allotype but not the "b" allotype. When AM14 BCR transgenic mice are bred onto the Balb/c background ("a" allotype), B cells specific for IgG2a can be found in the periphery, serum antibodies are present, and these cells can be activated by appropriate stimuli. Such responses are also found when AM14 BCR mice are bred onto a C57Bl/6 ("b" allotype) background, suggesting that presence of autoantigen does not affect the activation capacity of these cells (Hannum, Ni, Haberman, Weigert, & Shlomchik, 1996; Shlomchik, Zharhary, Saunders, Camper, & Weigert, 1993). The low affinity of this receptor for antigen is thought to enable this kind of escape from deletion in the bone marrow, as high-affinity, IgG2a-specific B cells from the 20.8.3 model undergo clonal deletion and are absent from the periphery (Wang & Shlomchik, 1997).

4.3. B-cell selection in the periphery and activation

Upon reaching the immature B-cell stage checkpoint, B cells that lack affinity for self altogether, or those with low-affinity self-reactivity, exit the bone marrow as transitional B cells. Cells at the transitional 1 (T1) stage give rise to transitional follicular B cells through upregulation of IgD and CD23 (T2 B cells). These cells are thought to be the precursors of mature follicular (FO) B cells in the periphery upon downregulation of IgM. T2 B cells can also give rise to a third type of transitional (T3) B cell thought to be the precursor of marginal zone B cells, which express high levels of CD21 and recirculate only in the spleen; it is through downregulation of CD23 that T3 B cells become marginal zone B cells. Marginal zone B cells, as well as their transitional precursors, are found only in the spleen (Pillai, Cariappa, & Moran, 2005). Functionally, marginal zone B cells respond promptly to blood-borne pathogens in a T cell-independent manner (Martin, Oliver, & Kearney, 2001; Oliver, Martin, Gartland, Carter, & Kearney, 1997). Between the transitional and follicular stages, a secondary checkpoint occurs as the percentage of autoreactive B cells decreases from ~50% of newly emergent B cells to ~20% of circulating B cells (Wardemann et al., 2003). This checkpoint likely occurs as a result of interaction of newly emergent B cells with self-antigens absent in the bone marrow (Meffre & Wardemann, 2008).

Follicular B cells are typically activated through T-cell help in the course of a germinal center reaction. During antigen exposure, B cells that have interacted with antigens through the BCR travel to secondary lymphoid organs to receive signals from T helper cells and continue their differentiation program. Activated B cells that interact with cognate T cells in the T-cell–B-cell border are thus selected to enter the germinal center reaction. A subset of activated B cells secretes abundant IgM, but it is short-lived and fails to enter germinal centers.

Germinal centers are characterized by two anatomically and functionally distinct compartments, the "dark zone" and the "light zone." B cells first enter the dark zone, where B cells divide rapidly and undergo somatic hypermutation (SHM) in their BCR variable regions to produce B cells with altered affinity for antigen, where only B cells with increased affinity receive the requisite survival signals. Cells that have undergone affinity maturation travel to the light zone, where follicular dendritic cells interspersed with entering B cells display antigens on the surface for selection of high-affinity binders. In the light zone, B cells can undergo CSR to produce antibodies of identical specificity but of different isotypes. B cells that fail to produce better binders die by apoptosis. B cells that bind antigens with improved affinity exit the germinal center as antibody-producing cells or memory B cells (Victora & Nussenzweig, 2012). The majority of B-cell lymphomas in children and adults are derived from germinal center B cells (Fig. 1.1). Aberrant SHM can cause gene translocations that promote malignant transformation of B cells (Nussenzweig & Nussenzweig, 2010; Pasqualucci et al., 2001). Moreover, the presence of autoreactive B cells bearing IgG suggests that SHM can introduce self-reactive variants that can give rise to autoimmune phenotypes (Tiller et al., 2007) (Fig. 1.1).

How does breakdown of tolerance occur in the periphery? SLE is a multifactorial disease that is likely triggered by a set of events that promote activation of both innate and adaptive responses to otherwise innocuous self-molecules. How breakdown of tolerance by self-molecules occurs is still not fully understood, but access to previously inaccessible epitopes, as may occur upon deficient clearance of apoptotic constituents, may be required for initiation of disease. Genetic defects in proteins involved in apoptotic debris clearance and DNA degradation have been linked to SLE (Rullo & Tsao, 2013). In mice with normal (non-autoimmune prone) genetic backgrounds where B cells bear transgenic BCRs specific for dsDNA, Smith (Sm) ribonucleoprotein antigen or IgG, self-reactive B cells are either anergic or clonally ignorant. However, when crossed to autoimmune-prone backgrounds, such as the MRL.Fas^{lpr} background, anergic and clonally ignorant self-reactive B cells are activated and enter the germinal center reaction in follicles or undergo activation and SHM in extrafollicular sites and secrete autoantibodies (Mandik-Nayak et al., 1999; Santulli-Marotto, Qian,

Ferguson, & Clarke, 2001; Wang & Shlomchik, 1999; William, Euler, Christensen, & Shlomchik, 2002). What determines whether a mouse is prone to autoimmunity? Some congenic strains spontaneously develop lupus, likely through a combination of inputs from several genes or pathways. Examples include the F1 from crosses between New Zealand black and white (NZB/NZW F1), the MRL.Fas^{lpr}, and the BXSB strains. These extensively studied mouse models typically show signs of disease that mimic those found in SLE patients such as enhanced titers of antinuclear antibodies (ANA), splenomegaly and lymphadenopathy (enlarged spleens and lymph nodes, respectively), proteinuria, and glomerulonephritis (Rullo & Tsao, 2013). Genetically manipulated mice may also exhibit such signs if the genes altered play important roles in disease pathogenesis. Mutations in BCR signaling regulation pathways that can lead to autoimmune-prone phenotype will be discussed in the next section but before we will introduce the BCR pathways that lead to B-cell activation by antigen.

5. BCR ACTIVATION BY ANTIGEN

5.1. Early BCR events upon antigen binding

Naive B cells circulate in the lymph and secondary organs until they encounter antigen in the B-cell follicles. Because antigen-independent interactions of the BCR are required for survival, as inferred from the rapid decay of B cells upon conditional ablation of BCR expression (Lam, Kuhn, & Rajewsky, 1997), BCR interaction with an antigen should overcome the "survival" threshold to achieve full B-cell activation.

The requirements for B-cell activation in terms of antigen valency continue to be a matter of debate. In lieu of the importance of antigen avidity and affinity required for B-cell selection and activation, it is important to revisit our current understanding of the valency requirements for B-cell activation. The crosslinking model was first proposed in the 1960s to explain B-cell activation, based on the ability of F(ab′)2 anti-IgM to promote B-cell proliferation and the lack of responses to monomeric Fab (Woodruff, Reid, & James, 1967). This model suggests that polyvalent antigens can bring BCR molecules together in close apposition, or "crosslink," and are best at activation. However, not all antigens are spatially organized such that several of the B-cell epitopes occur in close proximity, as would be the case for a virus particle or a bacterial surface, both of which can carry highly repetitive structures. Most soluble protein antigens bear single epitopes that can still trigger an immune response specific for that epitope.

Either there must exist some measures of aggregation, for example, by presentation of multiple copies of one and the same protein antigen on a surface, such as the plasma membrane of a dendritic cell, or alternative models for activation must apply. The dissociation–association model was first proposed by Reth and colleagues. Their model proposes that the BCR exists in a closed, inactivated state, present as autoinhibited oligomers in B cells at rest (Yang & Reth, 2010b). It is only upon binding of antigen that oligomers dissociate, for example, as a result of conformational changes in Igα and Igβ or by antigen serving as a wedge that creates distance between adjacent BCRs. These BCR monomers, now in "open" conformation, are relieved from further inhibitory constraints and can activate signaling. This model better explains B-cell activation by monovalent antigens (Yang & Reth, 2010a). Monovalent engagement of the BCR allows internalization of bound antigen (Avalos et al., 2014; Kim et al., 2006), but antigen presentation is more efficient for multivalent antigen capture (Kim et al., 2006).

The cytoskeleton associates with the BCR complex: the group of Batista demonstrated its essential role in activation of BCR signaling. BCR diffusion in the plane of the membrane depends on interactions with the underlying actin cytoskeleton: diffusion of resting BCRs is restricted, while in anti-IgM-activated B cells, BCRs diffuse more freely (Treanor et al., 2010). In fact, disruption of the cytoskeleton by treatment with latrunculin A or cytochalasin D induced BCR signaling events in the absence of antigen (Treanor et al., 2010). Such enhanced BCR diffusion is consistent with the demonstration that B cells undergo an expansion and contraction response upon antigen activation, which likely serves to help capture more antigen (Fleire et al., 2006). Even at rest, B cells display "scanning behavior" through extrusion of pseudopodia, a behavior that suggests these cells are likely probing for antigen (Ana Avalos & Hidde Ploegh unpublished observations). Certain local regions of resting B cells may thus have increased BCR diffusion rate. By stochastic optical reconstruction microscopy (STORM), it was shown that BCRs occur in small clusters in resting cells, consistent with the model proposed by Reth, and these may be in dynamic equilibrium with monomers (Mattila et al., 2013). Pierce and coworkers demonstrated that BCR is present in monomers at the cell surface, which aggregate upon antigen encounter, as assessed by increased Förster energy resonance transfer (FRET) between donor and acceptor fluorophores associated with Igα and Igβ (Tolar, Sohn, & Pierce, 2005). These events depend on the constant region of IgM, as the Cμ4 domain is required (Tolar, Hanna, Krueger, & Pierce, 2009) (more thoroughly reviewed elsewhere in this volume). Thus, it appears that BCRs behave dynamically on the B-cell surface, aided by interactions with the cytoskeleton. Changes in BCR distribution and mobility likely depend on the affinity or avidity of their interactions with proteins and polysaccharides present in the extracellular matrix. Such interactions may be relevant for survival, tonic signals that, as we will explore in the succeeding text, could lead to malignancies if not properly regulated. Binding of antigen to BCR that exceeds a particular threshold leads to activation through a cascade of signaling events that determine a B cell's fate. In the succeeding text, we revisit in detail such events.

5.2. BCR signaling pathways in response to antigen

The signaling pathways activated upon antigen–BCR interaction have been thoroughly characterized and involve activation signaling cascades that culminate in activation of transcription factors that ultimately determine the proliferation and differentiation of B cells. Covered in more detail later, autoreactive and malignant B cells exploit such pathways to promote their expansion and survival. As explained before, the μ chain that is part of the pre-BCR and becomes an integral subunit of the BCR on newly generated B cells lacks a cytoplasmic domain capable of extensive cytoplasmic interactions. Signaling through the pre-BCR and mature BCR depends on their noncovalent interaction with Igα and Igβ, which contain ITAMs in the cytoplasmic domain (Hombach, Tsubata, Leclercq, Stappert, & Reth, 1990). ITAM motifs have consensus sequences YxxI/Lx6-12YxxI/L where tyrosine residues are phosphorylated upon activation. Upon antigen binding, the Src tyrosine kinase Lyn, which is associated with the plasma membrane through acylation, phosphorylates the ITAMs, thus generating docking sites for proteins containing Src homology (SH2) domains. The SH2 domains of the Src tyrosine kinase Syk mediate recruitment to the ITAMs, together with Lyn, which amplifies early signals (Kurosaki, Shinohara, & Baba, 2010). BCR signaling also induces activation of phosphatidylinositol-3-kinase (PI3K) through recruitment to the surface receptor CD19. CD19 contains cytoplasmic domains with consensus sequence Y–X–X–M where again tyrosines are the targets for phosphorylation by Lyn. These phosphorylation events lead to recruitment of the p85 subunit of PI3K, which contains SH2 domains. B-cell adaptor for PI3K (BCAP) also recruits PI3K, which contains Y–X–X–M domains that—upon phosphorylation—can bind an SH2 motif in the p85 subunit of PI3K. PI3K phosphorylates phophatidylinositol-4,5-biphosphate (PIP_2) to phophatidylinositol-3,4,5-triphosphate (PIP_3). Generation of PIP_3 by PI3K promotes the recruitment of several pleckstrin-homology (PH)-

containing molecules, including the Src protein kinase Btk, which recruits and phosphorylates tyrosines in the adaptor protein BLNK to serve as docking sites for the SH2 domains of phospholipase C-gamma 2 (PLCγ2). BLNK, also known as SLP-76 and BASH, serves as a scaffolding platform from which several activation pathways emerge. PLCγ2 also contains PH domains to interact with PIP_3 and generates diacylglycerol (DAG) and inositol-1,4,5,-triphosphate (IP_3). DAG release induces activation of protein kinase C (PKC), which in turn mediates activation of mitogen-activated protein kinases (MAPK) extracellular signal-regulated kinase (ERK), c-Jun NH2 terminal kinase (JNK/SAPK), and p38. These kinases phosphorylate and activate different sets of transcription factors in cascade fashion: phosphorylated ERK activates Elk-1 and Myc, JNK/SAPK activates c-Jun and ATF-2, and p38 triggers ATF-2 and MAX (Dal Porto et al., 2004).

Binding of IP_3 to IP_3 receptors on the endoplasmic reticulum (ER) promotes the rapid release of calcium from the ER lumen into the cytoplasm, which in turn activates entry of extracellular calcium through store-operated calcium (SOC) channels (Kurosaki et al., 2010). Increase in cytosolic calcium activates the transcription factor nuclear factor of activated T cells (NFAT) via calcium–calmodulin interactions. Association of the guanine exchange factor Vav with BLNK also activates Rac and causes cytoskeletal alterations; the nucleotide exchange factors Sos and RasGRP3 activate Ras, which modulates the ERK pathway. BLNK also recruits Nck, a cytoskeleton-binding protein essential for BCR signaling through recruitment of PI3K to BCAP (Castello et al., 2013).

PI3K phosphorylation of PIP_2 to PIP_3 promotes recruitment of the serine/threonine kinase Akt through its PH domain, which mediates survival through regulation of expression of proteins involved in apoptosis as well as transcription factors, including members of the forkhead box O family (Hombach et al., 1990), E2F, NFκB and CREB (Dal Porto et al., 2004).

BCR ligation also activates IκB kinase (IKK) complex, which consists of three subunits (α, β, and γ). The NFκB inhibitor IκB is associated with NFκB subunits p105, p100, RelA, RelB, and c-Rel in the cytoplasm, and its removal by degradation requires activation of IKK. The released NFκB subunits then enter the nucleus and activate gene expression. Upon BCR activation, the generation of DAG and calcium activates protein kinase C-β (PKCβ), which likewise activates NFκB through the formation of a protein complex composed of the adaptor molecules Card11 (also known as CARMA), Malt1, and Bcl10 (Kurosaki et al., 2010). Card11 is essential for BCR- and PKC-induced proliferation and cytokine production. In Card11-defective mice, BCR activation does not result in NFκB or JNK activation. In addition, Card11 plays a role in JNK activation and B-cell proliferation in response to TLR4 stimuli (Hara et al., 2003; Jun et al., 2003).

The BCR transmits signals immediately upon antigen binding; bound antigen is internalized by means of clathrin-mediated endocytosis (Stoddart et al., 2002). Internalized antigen is then degraded by proteases to generate the peptides that are loaded onto class II MHC products for presentation to T helper cells to aid in B-cell activation and further differentiation (Bryant & Ploegh, 2004). As antigen is brought in, phosphorylated Lyn and Syk remain associated with transferrin receptor-positive (TfR+) compartments and are dephosphorylated before entering LAMP1+ compartments. Later, effectors of signaling such as the MAPK Erk, p38, and JNK are present first in TfR+ compartments and then accumulate in multivesicular structures, the site of antigen proteolysis and loading onto MHC II compartments. Proper signaling requires internalization: treatment with the dynamin inhibitor dynasore, which blocks clathrin-mediated internalization, imposes several defects in kinase pathways usually engaged upon B-cell activation (Chaturvedi, Martz, Dorward, Waisberg, & Pierce, 2011).

5.3. Regulation of BCR signaling

BCR signals are transduced within seconds, and positive regulators such as CD19 and negative regulators of these signals such as CD22 and FcγRIIB either expand or dampen them. CD19 is a 95-KDa glycoprotein that is upregulated at the pre-B-cell stage and remains on the B-cell surface until B cells differentiate into plasma cells. The cytoplasmic domain of CD19 is phosphorylated by Lyn and through recruitment of PI3K amplifies the signals emanating from antigen binding to BCR. In the absence of CD19, B cells are unable to respond to membrane-bound antigen but can sense soluble antigen in a comparable manner to normal, WT B cells (Depoil et al., 2008). Association of CD19 with surface CD21 (complement receptor type 2 that binds C3 complement components), the tetraspanin CD81, and leu13 molecules (Tedder, Inaoki, & Sato, 1997) provides a means of connecting the BCR complex with the complement receptors in response to antigens coupled to C3d. In fact, BCR responses to membrane-bound antigen require CD81, demonstrating that CD19–CD81 complexes are important for these responses (Mattila et al., 2013).

CD22 is a member of the sialic acid-binding immunoglobulin-like lectin (Siglec) family that binds α2,6-linked sialic acid residues. Sialic acids are common in vertebrates but are largely absent in bacteria. CD22 contains four immunoreceptor tyrosine-based inhibition motifs (ITIMs) with the consensus sequence S/I/V/LxYxxI/V/L. Binding of sialic acids to CD22 upon BCR interaction with an antigen that contains α2,6-linked sialic acids leads to phosphorylation of ITIMs and recruitment of phosphatases that shut down signals initiated by phosphorylation. This mechanism helps regulate self-antigen activation signals and promotes tolerance. Because B cells display a large amount of sialic acids at their surface, most sialic acid-binding sites in CD22 are likely occluded by occupancy by sialic acids in cis. This level of occupancy may set a regulatory threshold that a BCR-bound antigen must overcome. Lyn also phosphorylates the ITIMs in CD22 in the course of BCR activation by antigen (Nitschke, Carsetti, Ocker, Kohler, & Lamers, 1997). CD22 exerts its inhibitory effects by recruitment of SH2-containing tyrosine phosphatase-1 (SHP-1), whose substrates include CD19, BLNK, and Vav, in addition to the calcium ATPase PMCA, which extrudes calcium from the cell (Dorner, Shock, & Smith, 2012). CD22 can also recruit the SH2 domain-containing inositol-5-phosphatase 1 (SHIP-1). Whereas SHP-1 dephosphorylates tyrosine residues in key BCR signaling components, SHIP1 is a lipid phosphatase that catalyzes the conversion of PIP_3 to PIP_2. PIP_2 can no longer support the binding of PH domain-containing proteins such as Btk and PLCγ2, further dampening BCR signals.

Fc gamma receptors (FcγR) recognize the Fc portion of IgG and play an essential role in the regulation of immune responses. FcγR can be subdivided into activating FcγR and inhibitory FcγR. B cells express the inhibitory FcγRIIB, which contains ITIM motifs and can thus downregulate BCR responses when B cells interact with immune complexes containing IgG. B cells do not express activating FcγR. Upon interaction with IgG, FcγRIIB activates SHIP-1 and thus decreases recruitment of BCR activators to the signalosome complex. SHIPs also recruit Dok-1, which downregulates the activity of ERK (Nimmerjahn, Bruhns, Horiuchi, & Ravetch, 2005).

With such complex signaling pathways and regulatory signals that govern B-cell activation through the BCR, it is not surprising that mutations in components of these pathways can lead to breakdown of tolerance. First, we outline the triggers that activate B cells in SLE models and later in B-cell malignancies.

6. BREAKDOWN OF TOLERANCE BY BCR SIGNALING COMPONENTS

Mice with targeted disruption of genes that encode signaling mediators and accessory receptors with regulatory function such as tyrosine kinases, CD19, CD22, and FcγRIIB develop phenotypes of autoimmune disease (Fig. 1.2).

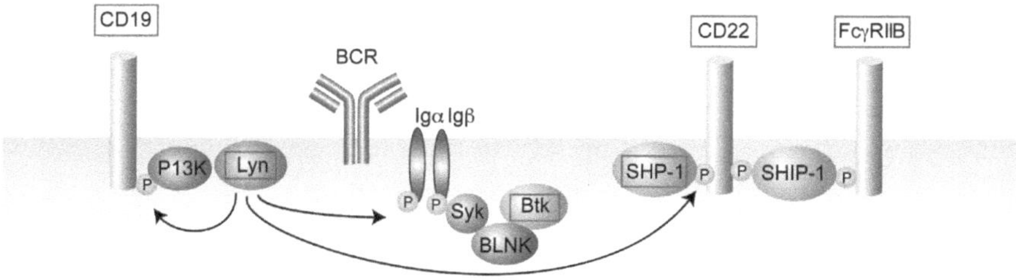

Figure 1.2 *Mutations in BCR signaling and regulator molecules give rise to mice exhibiting autoimmune phenotype.* Gene targeting either to upregulate expression of positive regulators (CD19) or to dampen signals from negative regulators (CD22, FcγRIIB, and SHP-1) can give rise to mice exhibiting features of human SLE. Mutation of tyrosine kinase Btk prevents autoimmune phenotypes, while the phenotype exerted by Lyn mutants is more complex due to its dual role as activator and suppressor of BCR responses. (See the color plate.)

6.1. Positive regulators: CD19 and tyrosine kinases

Targeted disruption of CD19 leads to a decrease in the mature B-cell pool in spleen and lymph nodes (~50% reduction). CD19−/− B cells display hyporesponsiveness to BCR signals and LPS and reduced levels of circulating immunoglobulins upon immunization with T cell-dependent antigens (Engel et al., 1995; Rickert, Rajewsky, & Roes, 1995). Conversely, transgenic CD19 overexpression leads to enhanced Ig secretion and spontaneous production of autoantibodies specific for ssDNA and IgG (Engel et al., 1995; Sato, Ono, Steeber, Pisetsky, & Tedder, 1996). In humans CD19 deficiency causes hypogammaglobulinemia with reduced numbers of CD27- and CD5-positive B cells (van Zelm et al., 2006).

Lyn is involved in B-cell selection and initiation of BCR responses. Conversely, Lyn−/− mice have decreased subsets of B cells in the periphery and altered responses to antigen. Moreover, these mice have increased titers of autoantibodies in the periphery and exhibit inflammation in the kidneys and immune complex deposition, possibly a reflection of defective selection checkpoints in development (Hibbs et al., 1995; Nishizumi et al., 1995). Lyn is also involved in dampening of responses to antigen and mice that upregulate Lyn have a complex phenotype with altered B-cell subsets in the periphery that can lead to autoimmune phenotype such as increased titers of IgM and circulating IgG, autoantibodies, and renal pathology due to immune complex deposition (Hibbs et al., 2002). Single-nucleotide polymorphisms have been associated with SLE patients of European descent (Lu et al., 2009), and patients with SLE exhibit low levels of Lyn and abnormal signaling that lead to higher proliferative activity and anti-DNA antibody production (Flores-Borja, Kabouridis, Jury, Isenberg, & Mageed, 2005). Thus, it appears that even though Lyn exerts both activation and regulatory roles, its role in regulation may be more prevalent than its role for activation, possibly due to modulation of association of SHP-1 with CD22 and the presence of other, likely redundant, tyrosine kinases that can exert positive roles (Fyn, Lck, and Blk) (Flaswinkel & Reth, 1994).

X-linked agammaglobulinemia (XLA) in humans and X-linked immunodeficiency (Xid) in mice are caused by a block in B-cell differentiation as a result of a Btk mutation. Xid mice crossed to autoimmune-prone models show ameliorated disease phenotypes (Seldin et al., 1987; Steinberg, Smathers, Frederiksen, & Steinberg, 1982). Xid mice crossed to transgenic AM14 (IgG-specific) or 56R (anti-DNA-specific) cannot respond to BCR signals alone but can do so upon coengagement of Toll-like receptor (TLR) 9 and the BCR, suggesting that Btk leads to a breakdown of tolerance. However, for this event to occur, TLR9 engagement is required (Halcomb, Musuka, Gutierrez, Wright, & Satterthwaite, 2008; Nundel, Busto, Debatis, & Marshak-Rothstein, 2013). We will discuss in the succeeding text how TLRs and BCRs synergize to activate autoreactive B cells. While a direct link with SLE has not been established, Btk is an important therapeutic target due to its central role in establishing human B-cell tolerance (Paz & Tsokos, 2013).

6.2. Negative regulators: CD22 and FcγRIIB

CD22−/− mice show normal B-cell development. Mature CD22−/− B cells exhibit heightened calcium responses, proliferation, and CD86 upregulation to BCR ligation using anti-μ antibody and increased responses to T cell-dependent antigens. In the absence of immunization, young CD22−/− mice show increased titers of IgM with dsDNA specificities (O'Keefe, Williams, Davies, & Neuberger, 1996). As the mice age, they develop somatically hypermutated high-affinity and class-switched anti-dsDNA and anticardiolipin IgGs (O'Keefe, Williams, Batista, & Neuberger, 1999). The autoimmune phenotype of CD22−/− mice appears to be dependent on genetic background of mice as CD22−/− mice in the C57Bl/6 background do not produce autoantibodies (Nitschke et al., 1997) yet mice in a mixed 129 × C57Bl/6 background do (O'Keefe et al., 1999, 1996). This dependency is also seen in FcγRIIB knockout mice, as it will be discussed in the succeeding text, and correlates with the heterogeneity found in autoimmune manifestations of patients with autoimmune disease.

As expected, the phenotype exhibited by the CD22−/− mice is mimicked by that found in SHP-1-deficient viable motheaten (*me*^{v}) mice (Cyster & Goodnow, 1995). SLE is associated with single-nucleotide polymorphisms in the gene encoding the integrin CD11b (Nath et al., 2008). CD11b, a marker for myeloid cells, is implicated in B-cell tolerance. Disruption in the *cd11b* gene in transgenic mice specific for the snRNP, which recognize the Smith (Sm) RNP antigen, leads to enhanced proliferative capacity and survival of B cells, and these responses are exclusive of BCR engagement and refractory to TLR stimulation (Ding et al., 2013). Immunization with apoptotic cells (thought to bear snRNP antigen) leads to an increase in autoantibody titers and Ig deposition in the kidneys. The contribution of CD11b appears to be dependent on binding to CD22 (Ding et al., 2013). Genetic ablation of CD22 promotes not only autoreactive B-cell activation but also B-cell neoplasia. Deletion of most of the cytoplasmic domain of CD22 (CD22ΔE12) has been seen in infant B-ALL and was associated with very aggressive growth of leukemia cells. Forced expression of mutated CD22 in transgenic mice induced hyperplasia of B cells and precursor B cells (Uckun, Goodman, Ma, Dibirdik, & Qazi, 2010). The same deletion was detected in the bone marrow of children with therapy-refractory ALL (Ma et al., 2012). CD22ΔE12 may therefore be a biomarker for very aggressive forms of B-ALL in infants and children.

C57Bl/6 mice with a targeted mutation in the FcγRIIB gene show decreased survival and display splenomegaly, glomerulonephritis, and IgG-IC deposits in the kidney, all features found in patients with SLE. FcγRIIB−/− mice also display elevated serum titers of antichromatin and antinuclear antibodies. This phenotype is not observed in mice on the Balb/c background (Bolland & Ravetch, 2000). The key role of FcγRIIB in maintenance of B-cell tolerance is inferred from experiments where reconstitution of lupus-prone mouse strains such as NZM 2410 and BXSB with bone marrow cells that were transduced with FcγRIIB yielded mice with significantly decreased SLE-like pathology (McGaha, Sorrentino, & Ravetch, 2005). In humans, a single-nucleotide polymorphism (SNP) in the *FCGR2B* gene that exchanges an isoleucine for a threonine at position 232 showed a significant association with susceptibility to SLE in Caucasians and Southeast Asians (Willcocks et al., 2010). SLE patients showed decreased FcγRIIB expression in memory B cells and plasmablasts, with a concomitant enhanced B-cell activation upon BCR engagement (Mackay et al., 2006). Thus, FcγRIIB is a central regulator of B-cell tolerance in the context of autoimmunity.

7. ABERRANT SIGNALING EVENTS DOWNSTREAM OF THE BCR AND NEOPLASIA

In general, aberrant signaling downstream of the BCR in neoplastic B cells is a result of increased BCR activation via either Igα/Igβ mutations or antigen binding, but not because of mutations in the individual downstream signaling molecules. Constitutive NFκB activation is characteristic for ABC-type DLBCL, with NFκB inhibition resulting in death of lymphoma cells (Davis, Brown, Siebenlist, & Staudt, 2001). Based on RNA interference, activation of NFκB depended on Card11 function (Ngo et al., 2006). Card11 mutations were identified in 10% of ABC-type DLBCL cases. These mutations activate the Card11/Bcl10/Malt1 complex with concomitant NFκB activation, improving survival of lymphoma cells (Lenz, Davis, et al., 2008) (Fig. 1.3). The Card11 mutation is the only mutation identified so far in the BCR signaling pathway. The components of the BCR signaling cascade have taken center stage in B-cell lymphoma and leukemia research because their selective blockade offers new therapeutic opportunities. Dasatinib is FDA-approved for targeting the ABL kinase, which is activated by the BCR–ABL translocation *t*(9;22) in chronic

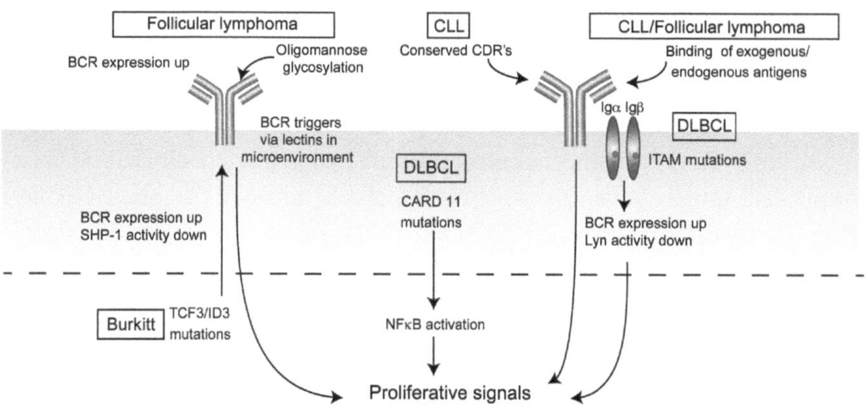

Figure 1.3 *The BCR plays a central role in lymphoma and leukemia pathogeneses.* Aberrant BCR signaling contributes to enhanced proliferative signals that promote the development of B-cell leukemia and lymphoma on different levels. BCR activity may be enhanced by binding of endogenous or exogenous antigens and via conserved CDRs. BCR expression may also be increased due to mutations in Igα/Igβ or the transcription factor TCF3. Changes in BCR glycosylation may induce BCR signaling via binding to lectins in the lymphoma microenvironment. Card11 mutations can enhance NFκB activation. (See the color plate.)

myeloid leukemia. It also inhibits Btk and Src family kinases (Hantschel et al., 2007) and has been used in a phase II trial to treat CLL (Amrein et al., 2011). Ibrutinib is a more specific Btk inhibitor. In a phase I study of adult patients with relapsed and refractory B-NHL and CLL, a complete or partial response was observed in 40% of participants (Advani et al., 2013). Ibrutinib is currently being tested in a phase II trial in DLBCL patients (relapsed and refractory). Other inhibitors tested in early clinical trials target Syk, PI3K, Src family kinases, and Torc1 (reviewed by Young and Staudt (2013)).

8. TONIC BCR SIGNALING

Antigen-independent BCR and pre-BCR signaling is termed tonic signaling (Monroe, 2006). The concept was first proposed based on the findings that the pre-BCR signals even though it does not contain a conventional ligand-binding site (Shaffer & Schlissel, 1997). BCR expression is required to maintain peripheral B-cell pools even though there is no direct evidence of antigen binding (Lam et al., 1997). Tonic signals are transmitted via the phosphorylated ITAM motifs of Igα and Igβ. Monroe suggested that tonic BCR signaling is a stochastic process and that ligand-dependent BCR signaling and ligand-independent BCR signaling are "opposite ends of a continuum." In this concept, tonic signaling is not induced and not localized to specific membrane regions. The tonic signal is transient and can be stabilized and increased by antigen-dependent signaling events. As antigen-dependent signaling increases, tonic signals will decrease (Monroe, 2006).

Acute B-ALL is the most common B-cell malignancy of childhood. Leukemic cells usually have a pro- or pre-B-cell phenotype. The pre-BCR and its downstream signaling molecules like BLNK function as tumor suppressors by inhibiting kinases like JAK3 (Nakayama et al., 2009). JAK3 is essential for common γchain-dependent B lymphoid signaling. Patients with JAK3 mutations present with the clinical picture of autosomal recessive severe combined immunodeficiency (Russell et al., 1995). BCR tumor suppressor properties depend on the transcription factor Ikaros, required for progression of B-cell development (Georgopoulos et al., 1994). Ikaros-deficient mice show a block in B-cell development at an aberrant pro-B-cell stage characterized by increased cell adhesion and loss of pre-BCR signaling (Schwickert et al., 2014). In mice, loss of Ikaros function contributes to B-cell leukemogenesis (Joshi et al., 2014). In pediatric patients, Ikaros mutations—including those that affect the DNA binding domain—are associated with a poor-prognosis B-ALL (Mullighan et al., 2009).

The BCR and its signaling components are crucial for survival of activated B cell-like (ABC)-type DLBCL. Knockdown of BCR components (IgH, Igκ, Igα, or Igβ) or of members of the signaling cascade (Syk, BLNK, PLCγδ, PI3K, and PKC) causes cell death. Bruton's kinase (Btk) activity is essential for ABC-type DLBCL survival (Davis et al., 2010). BCRs in ABC-type DLBCL cluster on the cell surface resembling BCR clustering in response to antigen exposure. In line with this observation, tyrosine phosphorylation was detected at the site of BCR clustering, a finding that shows BCR activation in ABC-type DLBCL. Some 20% of ABC-type DCBLs contain mutations in the ITAM of Igα or Igβ. These mutations increase BCR surface expression by preventing endocytosis. Mutated Igβ diminishes activity of Lyn, a positive and negative regulator of BCR signaling (Davis et al., 2010). Combined, these mutations increase the strength of the BCR signal (Fig. 1.3) and promote lymphoma cell proliferation.

Burkitt lymphomas express surface IgM, even though they are derived from germinal centers and should have undergone CSR (Klein, Klein, Ehlin-Henriksson, Rajewsky, & Kuppers, 1995). Antigen-dependent BCR signaling as a mechanism to support lymphoma growth has not been demonstrated. Most Burkitt lymphoma cell lines die after Igα knockdown and elimination of Syk has the same effect. Tonic BCR signaling therefore is essential for Burkitt lymphoma cell survival. It does not depend on Card11 or Btk and activates PI3K instead (Schmitz et al., 2012). In addition, 70% of Burkitt lymphoma samples analyzed in one study contained mutations of the transcription factor TCF3 or its negative regulator ID3 leading to increased TCF3 activity, which results in increased BCR expression and reduced expression of the BCR inhibiting the phosphatase SHP-1 (Schmitz et al., 2012) (Fig. 1.3). Tonic BCR signaling is crucial for survival of Burkitt lymphoma cells.

9. ANTIGEN-DRIVEN MALIGNANT B-CELL PERSISTENCE

As described in the preceding text, autoantigens drive the pathology of autoimmune diseases. In the same vein, persistence of neoplastic B cells can be propelled by autoantigens. Dameshek and Schwartz described similarities of autoimmune and leukemic lymphoproliferation and proposed the concept of autoantigen-dependent growth of leukemia cells in 1959 as a "speculative discussion" (Dameshek & Schwartz, 1959). Since then, it has become clear that antigen binding by BCRs expressed on neoplastic B cells supports the malignant phenotype and in some cases is even essential for survival of malignant cells. Both exogenous and endogenous antigens may be recognized. These may be found in the tumor microenvironment, which is receiving increasing attention as a therapeutic target (Burger, Ghia, Rosenwald, & Caligaris-Cappio, 2009). CLL can be classified according to the expression of germ line or somatically mutated BCR variable regions. The latter are associated with a more favorable prognosis (Damle et al., 1999; Hamblin, Davis, Gardiner, Oscier, & Stevenson, 1999). In CLL, the restricted and stereotypic BCR repertoire (Agathangelidis et al., 2012) is an argument for a BCR-driven selection process in which specific antigens stimulate expansion of malignant clones (Fig. 1.3). These clones share similar amino acids in the highly variable complementary determining region 3 (Ghiotto et al., 2004; Messmer et al., 2004). Most CLL cells express IgM and IgD. Surface Ig expression is a common feature of CLL and is lost only in rare cases, illustrating the importance of the BCR for CLL homeostasis. The Ig heavy-chain variable gene repertoire frequently uses *IGVH1-69*, *IGVH3-7*, *IGVH3-21*, and *IGVH4-34* (Fais et al., 1998). CLL cells utilizing *VH3-21* gene show restrictive *Vλ2-14* gene usage and homologous CDR3s, which strongly suggests recognition of some common epitope (Tobin et al., 2003). The germ-line BCRs of CLL clones often bind to autoantigens (Herve et al., 2005; Sthoeger et al., 1989). The discovery that CLL BCRs can bind to an invariant epitope of their own framework region 2 of VH domains to induce BCR signaling (Duhren-von Minden et al., 2012) highlights the complexity of autoimmunity as a mechanism that might contribute to CLL.

In SLE, apoptotic debris constituents may be the source of autoantigen driving pathology. BCR recognition of antigens derived from apoptotic cells and bacteria has been also demonstrated in CLL cells (Chu et al., 2010; Lanemo Myhrinder et al., 2008). Furthermore, human T-lymphotropic virus type 1 (HTLV type 1) has been linked to CLL development. The IgM isolated from CLL samples of HTLV-1-positive patients bound to HTLV-1-expressing cells and to the HTLV-I large envelope protein (Mann et al., 1987). The first exogenous antigen recognized by CLL BCRs carrying mutated IGHV genes has been identified: the cell wall component of yeast and filamentous fungi, β-(1,6)-glucan. CLL clones that express the respective BCRs proliferated in response to β-(1,6)-glucan (Hoogeboom et al., 2013). Commensals such as *Candida albicans* may thus promote growth of CLL cells. However, the *in vivo* consequences of BCR activation in CLL cells are not so easy to study. Upregulation of surface IgM on CLL clones after *in vitro* culture has been interpreted as a sign of BCR activation *in vivo* (Mockridge et al., 2007). The same is true for activation of the BCR signaling pathway in CLL samples (Muzio et al., 2008).

Not only CLL cells but also B-cell lymphomas use antigen-dependent BCR signaling to gain a proliferative advantage (Kuppers, 2005). To this end, mantle-cell lymphoma, DLBCL, follicular lymphoma, and Burkitt lymphoma express BCRs, mostly of the IgM isotype. In contrast, classical Hodgkin lymphomas have lost BCR expression (Schwering et al., 2003).

The biology of follicular lymphoma provides insight into the prominent role of antigen recognition by BCRs for lymphoma biology (Staudt, 2007). Follicular lymphoma cells continue to express IgM even under unfavorable circumstances, a trait not affected by the characteristic *t*(14;18) translocations that place Bcl-2 under control of IgH locus enhancers (Tsujimoto, Gorham, Cossman, Jaffe, & Croce, 1985; Vaandrager et al., 1998). These translocations occur in B-cell precursors in the bone marrow and may be caused by deregulation of the class-switching process. They are not sufficient to cause follicular lymphoma, but are considered to make affected cells vulnerable to other factors that cause malignant transformation (Kridel, Sehn, & Gascoyne, 2012). Autoreactive BCRs have been identified in follicular lymphoma patients (Dighiero et al., 1991), with 25% of follicular lymphoma BCR clones showing autoreactivity when tested by indirect immunofluorescence assay on HEp-2 cells (Fig. 1.3). One sample showed binding to myoferlin, a member of the ferlin family of proteins involved in membrane fusion, repair, and endocytosis. Myoferlin induced BCR signaling only in patient-derived lymphoma and not in normal B cells (Sachen et al., 2012). These findings—in addition to evidence of ongoing SHM—point to antigen selection as a central mechanism for clonal evolution in follicular lymphoma (Aarts et al., 2000; Bahler & Levy, 1992; Bahler, Zelenetz, Chen, & Levy, 1992). More recently, vimentin was identified as an autoantigen recognized by 20% of the follicular lymphoma samples analyzed. Vimentin is the major constituent of intermediate filaments in mesenchymal tissues (Franke, Schmid, Osborn, & Weber, 1978) and is thus ubiquitously expressed. The vimentin-reactive Igs did not share common CDR sequences and were not homologous. Vimentin was expressed in the T cell-rich regions of follicular lymphoma, underscoring the concept that the lymphoma microenvironment stimulates malignant B cells (Cha et al., 2013).

Follicular lymphoma cells have developed another mechanism to increase BCR signaling via interaction with the microenvironment. Their BCR antigen-binding site contains oligomannose glycans (Radcliffe et al., 2007) that can be crosslinked by mannose receptors expressed in the lymphoma microenvironment leading to BCR activation (Coelho et al., 2010). CLL surface IgM also carries high-mannose glycans, a finding that on the one hand implies antigen binding *in vivo* (Krysov et al., 2010) and on the other hand could have similar consequences for BCR activation by lectins (Fig. 1.3).

BCR triggering by infectious stimuli supports lymphoma growth in splenic marginal zone B-cell lymphomas. B-cell lymphomas can show an association with hepatitis C virus (HCV) infection (Marcucci & Mele, 2011). Direct binding of the lymphoma BCR to the viral envelope protein E2 has been demonstrated (Quinn et al., 2001). Treatment of HCV with IFNγ induced regression of HCV-associated splenic B-cell lymphoma (Hermine et al., 2002). Similarly, low-grade gastric lymphomas of mucosa-associated lymphoid tissue are associated with *Helicobacter pylori* infection and regress after antibiotic *Helicobacter* eradication (Wotherspoon et al., 1993). However, specificity of the gastric lymphoma BCRs for *H. pylori* antigens could not be demonstrated (Lenze et al., 2006). Instead, the expressed immunoglobulins are polyreactive (Craig et al., 2010). Autoimmune and chronic inflammatory diseases increase the risk for developing lymphoma, in support of the possibility that lymphoma is the result of BCR stimulation gone wrong (Smedby et al., 2006). Thus, a fundamental similarity between autoimmunity and B-cell malignancies is the maintenance of pathology by BCR interactions with self-antigens. Autoantigen can be found in the extracellular matrix, in the tumor microenvironment, or as a result of exposure to neoepitopes that neither type of B cell (autoreactive or neoplastic) has been exposed to before. Such novel molecules may bind with high affinity to these BCRs and trigger activation, which can result in antibody-mediated or neoplastic pathology.

10. THE ROLE OF TOLL-LIKE RECEPTORS IN BREAKDOWN OF B-CELL TOLERANCE IN AUTOIMMUNITY

Mature B cells in the periphery activate the proliferation and differentiation program in response to typically two signals: One is delivered by antigen via the BCR and the second signal is either delivered by the T cell (such as CD40L engagement of CD40 on the B cells) or by TLR signals (Goodnow, Vinuesa, Randall, Mackay, & Brink, 2010). Members of the TLR family are type I transmembrane proteins whose ligand recognition domain is composed of a number of leucine-rich repeats. TLR receptors are highly homologous between family members and recognize distinct molecular patterns in microbes. Depending on the localization of the microbial component they recognize, TLRs are located on the cell surface (surface TLRs) or intracellularly (endosomal TLRs). Endosomal TLRs recognize viral and bacterial DNA and RNA or their fragments, components that are released only upon degradation by professional phagocytes such as macrophages and dendritic cells. Endosomal TLR3 recognizes viral double-stranded RNA (dsRNA), while TLR7/8 recognizes single-stranded RNA (ssRNA); TLR9 senses unmethylated dsDNA from bacterial and viral origins that contains purine–purine–C–G–pyrimidine–pyrimidine consensus sequences (Krieg et al., 1995). Endosomal TLRs play an important role in recognition not only of pathogens but also of self-nucleic acids in autoimmune disorders.

Using the rheumatoid factor (IgG)-specific AM14 mouse model, immune complexes containing chromatin material mediated activation through convergence of BCR and TLR9 pathways (Leadbetter et al., 2002; Marshak-Rothstein, 2006; Viglianti et al., 2003). A similar situation was observed for immune complexes carrying TLR7 ligands (Lau et al., 2005). Thus, endosomal TLR antigens recognized through the BCR are internalized and delivered to TLR compartments where they start a proliferative response, and BCR engagement is required to bring IgM and TLRs to the same compartment (Chaturvedi, Dorward, & Pierce, 2008). An important question is how mammalian DNA mediates activation, considering that it is largely methylated and devoid of CpG motifs. Part of the answer seems to lie in the presence of CpG islands, portions of the genome often found in promoters, which are CG-rich and unmethylated. DNA fragments bearing such sequences were potent TLR9 activators, while CG-poor DNA was not stimulatory (Uccellini et al., 2008). FcγRIIB was also required to set a BCR threshold of activation by DNA and RNA immune complexes, as AM14 FcγRIIB−/− mice were more prone to respond to otherwise nonstimulatory DNA and RNA (Avalos, Uccellini, Lenert, Viglianti, & Marshak-Rothstein, 2010). In autoimmune-prone mice crossed to TLR7- and TLR9-deficient backgrounds, DNA-antibody specificities were dependent on the presence of functional TLR9 (Christensen et al., 2005; Ehlers, Fukuyama, McGaha, Aderem, & Ravetch, 2006), while RNA-related specificities were TLR7-dependent (Christensen et al., 2006). Targeted mutations of these TLRs elicited opposing effects, in that a TLR7 deficiency worsens disease, while a TLR9 deficiency prevents it through regulation of TLR7 (Christensen et al., 2006; Nickerson et al., 2010). In line with these findings, TLR8 and TLR9 regulate activity of TLR7 but do so in different cell types (Desnues et al., 2014). Cell-specific endosomal TLR9 behavior occurs as well, suggesting that cell-autonomous effects may be important for regulation of TLR activity (Avalos et al., 2013). How exactly one TLR could regulate the activity of another TLR remains to be established, but a possible mechanism may involve an intersection of the pathways of endosomal TLR trafficking.

Translocation of endosomal TLRs to their final destination is aided by the chaperone uncoordinated 93 homolog B1 (UNC93B1) (Lee, Avalos, & Ploegh, 2012). UNC93B1 is a polytopic protein with twelve transmembrane domains that is localized in the ER in resting cells and interacts with newly synthesized endosomal TLRs TLR3, TLR7, TLR9, and TLR13 through their transmembrane domains to deliver them to their final destination (Brinkmann et al., 2007; Kim, Brinkmann, Paquet, & Ploegh, 2008). The capacity to discriminate among TLRs is conferred at least in part by residues in the cytoplasmic domain of UNC93B1: a missense D34A mutation in the cytoplasmic domain of UNC93B1 causes enhanced TLR7 responses and dampened TLR9 responses, likely caused by augmented trafficking of TLR7 to endolysosomes (Fukui et al., 2009, 2011). UNC93B1 translocates TLR9 to endolysosomes through recruitment of adaptor protein 2 (AP-2), while TLR7 uses AP-4. The differential use of these adaptor proteins implies that TLR9 and TLR7 use different trafficking pathways (Lee et al., 2013).

11. CONTRIBUTIONS OF TLR SIGNALING TO ALTERED BCR ACTIVITY IN MALIGNANT B CELLS

While particular TLRs have not been linked to the pathogenesis of lymphomas and leukemias, the TLR signaling adaptor protein MyD88 has been implicated in a number of them. The MyD88 mutation L265P establishes a new oncogenic pathway in ABC-type DLBCL (Ngo et al., 2011) (Fig. 1.3). The mutation is localized in the MyD88 TIR domain and is a gain-of-function mutation that activates IRAK4 kinase activity. As described in the preceding text, tonic BCR signaling and Card11 mutations induce constitutive NFκB activation in ABC-type DLBCL that contribute to the poor cure rate of 40% in adult patients (Lenz, Davis, et al., 2008; Lenz, Wright, et al., 2008). In a subset of ABC-type DLBCL samples, JAK kinase signaling resulting in STAT3 activation was detected. The STAT3 activation further increased NFκB-mediated cell survival in DLBCL (Ding et al., 2008; Lam et al., 2008). To identify the molecular mechanisms of increased STAT3 activation, an RNA interference screen with ABC-type DLBCL lymphoma cell lines was performed. This screen showed that MyD88 and IRAK1 are required for cell line survival. Sequencing of the MyD88 gene showed that all cell lines contained the L265P mutation. When MyD88 was sequenced from 382 lymphoma biopsies, 30% of the ABC-type DLBCL samples contained the L265P mutant. Functional analysis demonstrated that it promotes DLBCL cell survival by inducing IRAK4 kinase activity, IRAK1 phosphorylation, NFκB signaling, and JAK kinase-mediated STAT3 activation. MyD88 and BCR signaling were shown to be nonredundant. Some DLBCL samples contained concomitant mutations in MyD88, Card11, and Igα/Igβ (Ngo et al., 2011). While multiple mutations of key signaling molecules in one lymphoma may explain a poorer prognosis in the first place, they can also become targets for new therapeutic approaches.

Recently, the effects of the L265P mutation on behavior of normal antigen-activated B cells were analyzed in mice (Wang, Jeelall, Beutler, Horikawa, & Goodnow, 2014). It initially caused TLR9-dependent proliferation and activation of NFκB signaling independently of exogenous TLR9 ligands. The activating and proliferative effects of the $\mathrm{MyD88}^{\mathrm{L265P}}$ mutation were self-limiting, through triggering of Tnfaip3. In collaboration with aberrant Bcl2 activity, the $\mathrm{MyD88}^{\mathrm{L265P}}$ mutation could prevent deletion of self-reactive B cells, allowing their accumulation. This shows that B cells use a multilayered approach to prevent breakdown of tolerance by endogenous TLR ligands. Bcl2 and TNFAIP3 are also frequently mutated in DLBCL (Ngo et al., 2011). These mutations increase uncontrolled signaling of mutated MyD88 in DLBCL cells and allow lymphoma cells to break self-tolerance. These data further support the concept that TLR and BCR pathways play a crucial role in tolerance induction and maintenance, and demonstrate that failure of proper signaling can promote uncontrolled B-cell proliferation paving the way for development of B-cell neoplasia and autoimmune diseases such as SLE.

12. SIMILARITIES AND DIFFERENCES BETWEEN THE STUDY OF ABERRANT B-CELL RECEPTOR SIGNALING IN AUTOIMMUNITY AND IN B-CELL MALIGNANCY

The B-cell receptor has been approached from different angles to study autoimmune pathology and B-cell malignancies. Such divergence may be traced to different concepts that underlie the etiology and clinical phenotype of autoimmunity and malignancy. Systemic autoimmune diseases are considered multifactorial resulting in multisystem disease; in contrast, B-cell neoplasias are thought to be B cell-intrinsic (caused by a malignant cell clone that has all necessary properties to cause malignancy (Hanahan & Weinberg, 2011)) resulting in disease symptoms that are due to neoplastic cells. As a result, the prominent experimental models differ:

a number of mouse models have been instrumental in our understanding of autoimmune disease. In the case of B-cell neoplasias, most research has focused in primary cells and cell lines derived from patients and analyzed *in vitro*, which are usually readily accessible. In B-cell neoplasia, mouse models have been most useful to study the molecular mechanisms of malignant transformation and clonal evolution and the effects of new therapies on lymphoma/leukemia growth. The preferred models are often humanized or immune-suppressed mice in order to allow xenotransplantation of human malignant B cells into mice. Despite these differences, recognition of autoantigens and BCR-sustained stimulation emerges as a common theme in supporting abnormal B-cell function in both autoimmunity and B-cell malignancy. Some forms of autoimmunity such as autoimmune lymphoproliferative syndrome (ALPS) can evolve gradually into lymphoma. Autoantigens in SLE have been identified, and the importance of self-antigens for growth and survival in B-cell malignancies has been described here. Pathology in autoimmune manifestations may involve several other cell types, including T cells and innate immune cells, activated by self-antigens that drive B-cell growth and that potentiate B-cell activation through production of cytokines in a positive feedback loop (Christensen & Shlomchik, 2007). The pathology of B-cell malignancies appears to be B cell-intrinsic. Whether other cells may contribute to their activation or prevalence is uncertain. However, the lymphoma/leukemia microenvironment is increasingly recognized for its contribution to development and homeostasis of malignant cells. Abnormal B cells can also utilize TLR pathways in both types of disease, and TLRs are important for maintenance of tolerance to endogenous constituents. Another common aspect shared by autoreactive and malignant B cells are mutations in or increased activity of key BCR signaling components such as Btk and Lyn; these can contribute to pathology. Therapies that target tyrosine kinases by small molecule inhibitors to treat both autoimmune manifestations and B-cell malignancies have shown efficacy in mouse models. Small molecule inhibitor therapies that target Btk have shown improved disease scores and IgG levels in the collagen-induced arthritis (CIA) mouse model. They have caused modest reductions in proteinuria and histopathology in MRL.Fas^{lpr} mice (Honigberg et al., 2010) and reductions in pathology in (NZBxW) F1 mice (Rankin et al., 2013). Consequently, these approaches are being considered for treatment of autoimmunity in man (Paz & Tsokos, 2013). B cell-depleting monoclonal antibodies (anti-CD20 and anti-CD22) have shown some success in the treatment of both these disorders, but their mechanism of action relies on the removal of harmful B cells. It is remarkable that despite the fundamentally different approaches that scientists have used to study autoimmune diseases and B-cell malignancies, several similarities could still be found. This suggests that perhaps, more parallels could be drawn should these disorders be studied in a more integrated manner, for example, by studying B-cell neoplasias as a "systems disease" instead of a B cell-intrinsic disease. Novel tools should be necessary for such undertaking. In addition, the emerging focus on biologics that aim to restore lost functions in B-cell disorders implies that we are in need to better understand the basis of B-cell activation in health and disease, and novel mouse models may aid in such endeavors.

13. TRANSNUCLEAR MICE: FILLING THE GAPS LEFT BY B-CELL TRANSGENESIS

13.1. Advantages of somatic cell nuclear transfer (SCNT)

While transgenesis and gene replacement technologies have had an enormous impact on our understanding of lymphocyte development, these methods were recently complemented by the application of SCNT as a means of obtaining a source of monoclonal lymphocytes of defined specificity. Earlier-generation BCR transgenics suffered from the drawback that neither the sites of transgene integration nor the levels and timing of BCR expression would faithfully mimic that of endogenous BCRs. Gene replacement strategies are more cumbersome in the design of the necessary targeting vectors but have the advantage that expression of the desired BCR is driven from the position of the orthologous locus. The number of such gene-targeted mice remains limited and often concerns specificities for haptens such as nitrophenol (NP) or highly repetitive structures such as DNA. Knockin mice with BCRs specific for myelin oligodendrocyte glycoprotein (MOG) and for HEL have found extensive use, but to our knowledge, the epitopes recognized by these BCRs cannot easily be copied through the use of synthetic peptides. Seen from this perspective, it is remarkable that until recently (Dougan et al., 2012), no mouse models existed with a BCR specific for ovalbumin, given the widespread use of TCR transgenic mice with CD4 and CD8 T cell-derived transgenic TCRs.

The process of SCNT consists of harvesting nuclei from a donor cell population and transplanting these nuclei into enucleated oocytes. Upon activation *in vitro* with strontium, the manipulated oocytes will differentiate into blastocysts. Although these SCNT blastocysts could be directly implanted into female recipients that will then carry the fetus to term, such direct cloning is not particularly efficient. Far more practical is the generation of an embryonic stem cell line from the SCNT-blastocyst inner cell mass. This provides a permanent source of cells that can be injected in multiple normal blastocysts to obtain chimeras and ultimately germ-line transmission of the V(D)J combinations carried by the donor lymphocyte (Kirak et al., 2010a; Kishigami et al., 2006). Importantly, this process is rapid, requiring approximately 6 weeks from isolation of donor nuclei to birth of chimeric mice.

Hochedlinger and Jaenisch first performed SCNT using donor nuclei from lymphocytes to rigorously establish the ability of a terminally differentiated cell to be reprogrammed into a totipotent stem cell (Hochedlinger & Jaenisch, 2002). The presence of the diagnostic V(D)J rearrangements in all somatic cells from mice born through this procedure is formal proof that such reprogramming is possible and did indeed occur. The Hochedlinger and Jaenisch experiments were performed with B and T cells of unknown specificity, since their goal was to demonstrate reprogramming and not necessarily the generation of new mouse models with antigen receptors of known specificity. The isolation of CD8 T cells of known specificity has been greatly facilitated through the development of MHC tetramer technology, now a standard tool for the T-cell immunologist. Production of class II MHC tetramers, while somewhat more challenging, is also feasible. These methods have provided ready access to even small populations of antigen-specific T cells at different stages of differentiation. Accordingly, we have succeeded in generating a sizable number of TCR transnuclear mice derived from CD8 T cells that are specific for *Toxoplasma gondii* antigens, for the tumor antigen TRP-1, and for a mouse gammaherpesvirus 68-derived antigen (Dougan, Dougan, et al., 2013; Kirak et al., 2010b; Sehrawat et al., 2012). It has been more difficult to isolate antigen-specific B cells. In naive mice, the frequency of antigen-specific B cells is too low to be practical for an SCNT experiment: some 200–300 cells are needed to ensure a reasonable likelihood of success, and we have not been able to isolate naive B cells of known antigen specificity from nonimmunized mice. We have therefore resorted to immunization to increase the frequency of antigen-specific B cells with the unavoidable outcome that their increased frequency in immune mice in many cases will be the result of B cells that underwent CSR and SHM. In this fashion, we generated mice by SCNT, using the nuclei form ovalbumin-specific B cells as the donor. Notwithstanding hyperimmunization, the ability to detect antigen-specific B cells by FACS using fluorescently labeled ovalbumin as probe proved challenging, but we succeeded in establishing the OB1 line of mice, which bear an ovalbumin-specific IgG1 (Dougan et al., 2012). In a similar fashion, we generated mice with an influenza virus hemagglutinin-specific IgG2b (Dougan, Ashour, et al., 2013). Both of these lines of mice share the interesting property that B-cell development is near normal. These observations suggest that pre-B-cell development can proceed normally even if the isotype of the pre-BCR heavy chain is not μ (Dougan, Ashour, et al., 2013; Dougan et al., 2012).

An additional benefit of the production of mice bearing homogeneous B-cell repertoires with already class-switched BCRs is the availability of models to study the process of CSR. The constant region of IgH in mice is composed of (in this order) μ, δ, γ3, γ1, γ2β, γ2α, ε, and α. Sequential switching has been demonstrated (Yoshida et al., 1990). It has been shown that sequential switching from IgM, to IgG1, to IgE enhances affinity of Ig for antigen, while IgM to IgE switching can also occur (Xiong, Dolpady, Wabl, Curotto de Lafaille, & Lafaille, 2012). Owing to SHM coincident with CSR, the limitation of using IgM B cells for these studies is the heterogeneity of the initial switched population (IgX from IgM), which can be translated to the second switch event (IgY from IgX). Moreover, the mechanisms that regulate sequential versus direct switching remain to be elucidated. Future studies will have to confirm that transnuclear mice bearing primary B-cell repertoires homogenous for antigen with already switched IgG can indeed undergo secondary CSR. The availability of the transnuclear B-cell models would overcome the initial heterogeneity of B-cell populations in mice bearing knockin IgM BCRs. Frequencies for accessibilities to downstream constant regions should thus be within reach. With the emerging role of the constant region of Ig as a modulator of B-cell responses to antigen (Casadevall & Janda, 2012; Tudor et al., 2012), models that can be reliably used to produce sequentially class-switched populations will expand our understanding of BCR responses to antigen. In addition, this knowledge will fill in the gaps in underexplored processes such as triggers of allergies and oral tolerance. With regard to autoimmunity and B-cell neoplasias, it will be possible to isolate how class-switched and memory B cells could contribute self-reactivities upon rechallenge with antigen or the process by which autoreactive B-cell specificities arise in the course of a GC reaction.

One can also study IgG BCR B cells to gain understanding of the BCR-intrinsic features that permit the prompt response of memory B cells to antigen rechallenge, and transnuclear mice provide such benefits. Memory

B cells might recognize neoepitopes or become autoreactive or even malignant B cells as a consequence of genetic alterations inherent to SHM. These aspects have been underexplored mostly due to the absence of adequate models to perform such studies.

13.2. Using transnuclear mice to study antigen valency and outcomes of B-cell activation

One particular aspect of B-cell activation for which the IgG1 BCR has proven useful is the relationship between antigen valency and BCR triggering. A limitation has been the absence of a suitable well-defined and antigen-specific model. In the HEL-specific BCR transgenic mouse model, chemical crosslinking of HEL was used to obtain monomeric, dimeric, trimeric, and tetrameric species (Kim et al., 2006). Monomeric HEL triggered early B-cell responses but failed to induce efficient antigen presentation (Kim et al., 2006). One confounding factor in these studies is the tendency of proteins to form aggregates, in both the presence and absence of serum. It is formally impossible to exclude the presence of higher-order structures even in the most rigorously purified and characterized monomeric preparations of a protein antigen, nor is it straightforward to eliminate adsorption of otherwise monomeric protein antigens to the surfaces of the reaction vessels in which such experiments are commonly done, as this would provide a multivalent, high-avidity surface for interaction with B cells. The use of the 3-nitro-4-hydroxy-5-iodo-phenylacetate (NIP) hapten-specific B1.8 model allowed the exploration of different peptides modified with one or more NIP haptens; here, monovalent antigen failed to induce any responses (Minguet, Dopfer, & Schamel, 2010), but the NIP-specific BCR is of comparatively modest affinity. The use of soluble monovalent antigens to trigger responses may be a poor substitute when membrane-bound monovalent antigen may be a more efficient means of delivering monovalent antigen in the two-dimensional space of the immune synapse (Batista, Iber, & Neuberger, 2001). The question stays as to whether the antigen-occupied BCR remains monovalent.

Using transnuclear OB1 mice bearing IgG BCR specific for ovalbumin (Dougan et al., 2012), we showed that monovalent minimal epitopes can activate B-cell responses (Avalos et al., 2014). The epitope required for recognition by serum from OB1 mice was identified by a series of overlapping peptides spanning the full ovalbumin sequence (Dougan et al., 2012). The OB1 model is unique in that as far as we know, this is the only BCR mouse model for which a linear epitope has been mapped. Moreover,

17-amino-acid-long peptides bearing the epitope are sufficient to trigger early BCR activation (Avalos et al., 2014). A reduction in the size of peptides bearing the epitope revealed an equivalent decrease in the activation signal, but this was not due to decreased binding (Avalos et al., 2014). Importantly, monovalent antigen triggered clustering of the BCR, suggesting that either conformational changes or interactions with the cytoskeleton as a result of a signal perceived in the variable region of BCR and transmitted to the constant region domains could lead to activation and aggregation. We found that only 5% of IgG was clustered on the cell surface of unstimulated B cells, suggesting that most BCR remains monomeric in these cells, consistent with results previously published for IgM (Tolar et al., 2005).

An 8-mer peptide bearing the OB1 epitope elicited a low-grade signal (as exemplified by a suboptimal flux of calcium upon BCR triggering). This signal blocked further activation by antigen or crosslinking agents such as anti-IgG or anti-kappa. This response was dependent on the strength of signal, since it was phenocopied by a very low concentration of 17-mer (70 × less) and lasted as long as the blocking agent was bound to the BCR (Avalos et al., 2014). BCRs that perceive a signal below threshold may prevent subsequent triggering as long as the low-stimulating antigen is present. Such behavior is consistent with that proposed for anergic B cells and may also apply as "tonic signaling" required for survival. Depending on the quality of signal (defined by not only the affinity for antigen but also other parameters such as the constants for association and dissociation and valency), the B-cell activation outcomes may differ.

Such a threshold-sensing mechanism is consistent with the different functions that BCR plays in B-cell development, selection, and activation. BCR activation outcomes may also result from the balance of positive and negative signals from other receptors that antagonize or synergize with BCR signaling. A balance between BCR-intrinsic factors, such as the quality of binding of antigen; BCR-extrinsic factors (positive and negative regulators of signaling); and the microenvironment may ultimately determine a "normal" development and behavior of B cells. Tipping of such balance may thus result in B cell-mediated disease.

14. CONCLUDING REMARKS

Despite the differences between etiology and clinical manifestations of autoimmune diseases and B-cell malignancies, there are common aspects that deserve further exploration as possible approaches to therapy. Thus, oncology and clinical immunology complement each other. For oncologists, the function of the BCR in initiating and sustaining malignant disease serves as an example of how malignant cells can interact with their host to get a proliferative advantage. Like SLE and other systemic disorders, malignancy may be more of a "system disease." We must therefore look for cross talk of malignant cells with other cells in the host that may, as for autoimmunity, contribute to the abnormal B-cell behavior. Defects in immune system may well act as cofactors that promote B-cell malignancy.

Breakdown of B-cell tolerance, while mostly dependent on antigen binding to BCR and hence BCR pathways, is attained by activation of and cross talk with other receptors such as TLRs. The aberrant proliferative capacity initially acquired by self-reactive and neoplastic B cells may depend on deregulated activation of NFκB as a result of TLR pathway activations. If the BCR fails to contain aberrant activation, then TLRs may be engaged and further expand such activation, sustained at a subsequent stage by a genetic contribution (oncogenes, mutations in molecules involved in apoptotic clearance pathways). The environment that surrounds the abnormally activated B cell (tumor environment, or other cells of the immune system that respond to TLR ligands and sustain B-cell activation) contributes as well. Thus, targeting the TLR pathways may be beneficial to limit the expansive ability of self-reactive or neoplastic B cells and perhaps of accessory cells or tissues that aid such abnormal activation. Future studies may well draw additional parallels between autoimmunity and neoplasia.

ACKNOWLEDGMENTS

We would like to thank Ann Marshak-Rothstein and Stephanie Dougan for critical reading of our manuscript and Tom DiCesare for artwork. H. P. was funded by NIH RO1 GM100518 and F. M. W. by the Dutch Cancer Foundation (KWF 2010-4628) and the WKZ foundation (WKZ fonds).

REFERENCES

Aarts, W. M., Bende, R. J., Steenbergen, E. J., Kluin, P. M., Ooms, E. C., Pals, S. T., et al. (2000). Variable heavy chain gene analysis of follicular lymphomas: Correlation between heavy chain isotype expression and somatic mutation load. *Blood*, *95*(9), 2922–2929.

Advani, R. H., Buggy, J. J., Sharman, J. P., Smith, S. M., Boyd, T. E., Grant, B., et al. (2013). Bruton tyrosine kinase inhibitor ibrutinib (PCI-32765) has significant activity in patients with relapsed/refractory B-cell malignancies. *Journal of Clinical Oncology: Official Journal of the American Society of Clinical Oncology*, *31*(1), 88–94. http://dx.doi.org/10.1200/JCO.2012.42.7906.

Agathangelidis, A., Darzentas, N., Hadzidimitriou, A., Brochet, X., Murray, F., Yan, X. J., et al. (2012). Stereotyped B-cell receptors in one-third of chronic lymphocytic leukemia:

A molecular classification with implications for targeted therapies. *Blood, 119*(19), 4467–4475. http://dx.doi.org/10.1182/blood-2011-11-393694.

Amrein, P. C., Attar, E. C., Takvorian, T., Hochberg, E. P., Ballen, K. K., Leahy, K. M., et al. (2011). Phase II study of dasatinib in relapsed or refractory chronic lymphocytic leukemia. *Clinical Cancer Research, 17*(9), 2977–2986. http://dx.doi.org/10.1158/1078-0432.CCR-10-2879.

Avalos, A. M., Bilate, A. M., Witte, M. D., Tai, A. K., He, J., Frushicheva, M. P., et al. (2014). Monovalent engagement of the BCR activates ovalbumin-specific transnuclear B cells. *The Journal of Experimental Medicine, 211*(2), 365–379. http://dx.doi.org/10.1084/jem.20131603.

Avalos, A. M., Kirak, O., Oelkers, J. M., Pils, M. C., Kim, Y. M., Ottinger, M., et al. (2013). Cell-specific TLR9 trafficking in primary APCs of transgenic TLR9-GFP mice. *Journal of Immunology, 190*(2), 695–702. http://dx.doi.org/10.4049/jimmunol.1202342.

Avalos, A. M., Uccellini, M. B., Lenert, P., Viglianti, G. A., & Marshak-Rothstein, A. (2010). FcgammaRIIB regulation of BCR/TLR-dependent autoreactive B-cell responses. *European Journal of Immunology, 40*(10), 2692–2698. http://dx.doi.org/10.1002/eji.200940184.

Bahler, D. W., & Levy, R. (1992). Clonal evolution of a follicular lymphoma: Evidence for antigen selection. *Proceedings of the National Academy of Sciences of the United States of America, 89*(15), 6770–6774.

Bahler, D. W., Zelenetz, A. D., Chen, T. T., & Levy, R. (1992). Antigen selection in human lymphomagenesis. *Cancer Research, 52*(19 Suppl), 5547s–5551s.

Batista, F. D., Iber, D., & Neuberger, M. S. (2001). B cells acquire antigen from target cells after synapse formation. *Nature, 411*(6836), 489–494. http://dx.doi.org/10.1038/35078099.

Bolland, S., & Ravetch, J. V. (2000). Spontaneous autoimmune disease in Fc(gamma)RIIB-deficient mice results from strain-specific epistasis. *Immunity, 13*(2), 277–285. http://dx.doi.org/10.1016/S1074-7613(00)00027-3.

Brinkmann, M. M., Spooner, E., Hoebe, K., Beutler, B., Ploegh, H. L., & Kim, Y. M. (2007). The interaction between the ER membrane protein UNC93B and TLR3, 7, and 9 is crucial for TLR signaling. *The Journal of Cell Biology, 177*(2), 265–275. http://dx.doi.org/10.1083/jcb.200612056.

Bryant, P., & Ploegh, H. (2004). Class II MHC peptide loading by the professionals. *Current Opinion in Immunology, 16*(1), 96–102.

Burger, J. A., Ghia, P., Rosenwald, A., & Caligaris-Cappio, F. (2009). The microenvironment in mature B-cell malignancies: A target for new treatment strategies. *Blood, 114*(16), 3367–3375. http://dx.doi.org/10.1182/blood-2009-06-225326.

Cambier, J. C., Gauld, S. B., Merrell, K. T., & Vilen, B. J. (2007). B-cell anergy: From transgenic models to naturally occurring anergic B cells? *Nature Reviews. Immunology*, 7(8), 633–643. http://dx.doi.org/10.1038/nri2133.

Casadevall, A., & Janda, A. (2012). Immunoglobulin isotype influences affinity and specificity. *Proceedings of the National Academy of Sciences of the United States of America, 109*(31), 12272–12273. http://dx.doi.org/10.1073/pnas.1209750109.

Castello, A., Gaya, M., Tucholski, J., Oellerich, T., Lu, K. H., Tafuri, A., et al. (2013). Nck-mediated recruitment of BCAP to the BCR regulates the PI(3)K-Akt pathway in B cells. *Nature Immunology, 14*(9), 966–975. http://dx.doi.org/10.1038/ni.2685.

Cha, S. C., Qin, H., Kannan, S., Rawal, S., Watkins, L. S., Baio, F. E., et al. (2013). Nonstereotyped lymphoma B cell receptors recognize vimentin as a shared autoantigen. *The Journal of Immunology, 190*(9), 4887–4898. http://dx.doi.org/10.4049/jimmunol.1300179.

Chaturvedi, A., Dorward, D., & Pierce, S. K. (2008). The B cell receptor governs the subcellular location of Toll-like receptor 9 leading to hyperresponses to DNA-containing antigens. *Immunity, 28*(6), 799–809. http://dx.doi.org/10.1016/j.immuni.2008.03.019.

Chaturvedi, A., Martz, R., Dorward, D., Waisberg, M., & Pierce, S. K. (2011). Endocytosed BCRs sequentially regulate MAPK and Akt signaling pathways from intracellular compartments. *Nature Immunology*, *12*(11), 1119–1126. http://dx.doi.org/10.1038/ni.2116.

Christensen, S. R., Kashgarian, M., Alexopoulou, L., Flavell, R. A., Akira, S., & Shlomchik, M. J. (2005). Toll-like receptor 9 controls anti-DNA autoantibody production in murine lupus. *The Journal of Experimental Medicine*, *202*(2), 321–331.

Christensen, S. R., & Shlomchik, M. J. (2007). Regulation of lupus-related autoantibody production and clinical disease by Toll-like receptors. *Seminars in Immunology*, *19*(1), 11–23. http://dx.doi.org/10.1016/j.smim.2006.12.005.

Christensen, S. R., Shupe, J., Nickerson, K., Kashgarian, M., Flavell, R. A., & Shlomchik, M. J. (2006). Toll-like receptor 7 and TLR9 dictate autoantibody specificity and have opposing inflammatory and regulatory roles in a murine model of lupus. *Immunity*, *25*(3), 417–428.

Chu, C. C., Catera, R., Zhang, L., Didier, S., Agagnina, B. M., Damle, R. N., et al. (2010). Many chronic lymphocytic leukemia antibodies recognize apoptotic cells with exposed nonmuscle myosin heavy chain IIA: Implications for patient outcome and cell of origin. *Blood*, *115*(19), 3907–3915. http://dx.doi.org/10.1182/blood-2009-09-244251.

Coelho, V., Krysov, S., Ghaemmaghami, A. M., Emara, M., Potter, K. N., Johnson, P., et al. (2010). Glycosylation of surface Ig creates a functional bridge between human follicular lymphoma and microenvironmental lectins. *Proceedings of the National Academy of Sciences of the United States of America*, *107*(43), 18587–18592. http://dx.doi.org/10.1073/pnas.1009388107.

Craig, V. J., Arnold, I., Gerke, C., Huynh, M. Q., Wundisch, T., Neubauer, A., et al. (2010). Gastric MALT lymphoma B cells express polyreactive, somatically mutated immunoglobulins. *Blood*, *115*(3), 581–591. http://dx.doi.org/10.1182/blood-2009-06-228015.

Cyster, J. G., & Goodnow, C. C. (1995). Protein tyrosine phosphatase 1C negatively regulates antigen receptor signaling in B lymphocytes and determines thresholds for negative selection. *Immunity*, *2*(1), 13–24.

Dal Porto, J. M., Gauld, S. B., Merrell, K. T., Mills, D., Pugh-Bernard, A. E., & Cambier, J. (2004). B cell antigen receptor signaling 101. *Molecular Immunology*, *41*(6–7), 599–613. http://dx.doi.org/10.1016/j.molimm.2004.04.008.

Dameshek, W., & Schwartz, R. S. (1959). Leukemia and auto-immunization- some possible relationships. *Blood*, *14*, 1151–1158.

Damle, R. N., Wasil, T., Fais, F., Ghiotto, F., Valetto, A., Allen, S. L., et al. (1999). Ig V gene mutation status and CD38 expression as novel prognostic indicators in chronic lymphocytic leukemia. *Blood*, *94*(6), 1840–1847.

Davis, R. E., Brown, K. D., Siebenlist, U., & Staudt, L. M. (2001). Constitutive nuclear factor kappaB activity is required for survival of activated B cell-like diffuse large B cell lymphoma cells. *The Journal of Experimental Medicine*, *194*(12), 1861–1874.

Davis, R. E., Ngo, V. N., Lenz, G., Tolar, P., Young, R. M., Romesser, P. B., et al. (2010). Chronic active B-cell-receptor signalling in diffuse large B-cell lymphoma. *Nature*, *463*(7277), 88–92. http://dx.doi.org/10.1038/nature08638.

Depoil, D., Fleire, S., Treanor, B. L., Weber, M., Harwood, N. E., Marchbank, K. L., et al. (2008). CD19 is essential for B cell activation by promoting B cell receptor-antigen microcluster formation in response to membrane-bound ligand. *Nature Immunology*, *9*(1), 63–72. http://dx.doi.org/10.1038/ni1547.

Desnues, B., Macedo, A. B., Roussel-Queval, A., Bonnardel, J., Henri, S., Demaria, O., et al. (2014). TLR8 on dendritic cells and TLR9 on B cells restrain TLR7-mediated spontaneous autoimmunity in C57BL/6 mice. *Proceedings of the National Academy of Sciences of the United States of America*, *111*(4), 1497–1502. http://dx.doi.org/10.1073/pnas.1314121111.

Dighiero, G., Hart, S., Lim, A., Borche, L., Levy, R., & Miller, R. A. (1991). Autoantibody activity of immunoglobulins isolated from B-cell follicular lymphomas. *Blood, 78*(3), 581–585.

Ding, C., Ma, Y., Chen, X., Liu, M., Cai, Y., Hu, X., et al. (2013). Integrin CD11b negatively regulates BCR signalling to maintain autoreactive B cell tolerance. *Nature Communications, 4*, 2813. http://dx.doi.org/10.1038/ncomms3813.

Ding, B. B., Yu, J. J., Yu, R. Y., Mendez, L. M., Shaknovich, R., Zhang, Y., et al. (2008). Constitutively activated STAT3 promotes cell proliferation and survival in the activated B-cell subtype of diffuse large B-cell lymphomas. *Blood, 111*(3), 1515–1523. http://dx.doi.org/10.1182/blood-2007-04-087734.

Dorner, T., Shock, A., & Smith, K. G. (2012). CD22 and autoimmune disease. *International Reviews of Immunology, 31*(5), 363–378. http://dx.doi.org/10.3109/08830185.2012.709890.

Dougan, S. K., Ashour, J., Karssemeijer, R. A., Popp, M. W., Avalos, A. M., Barisa, M., et al. (2013). Antigen-specific B-cell receptor sensitizes B cells to infection by influenza virus. *Nature, 503*(7476), 406–409. http://dx.doi.org/10.1038/nature12637.

Dougan, S. K., Dougan, M., Kim, J., Turner, J. A., Ogata, S., Cho, H. I., et al. (2013). Transnuclear TRP1-specific CD8 T cells with high or low affinity TCRs show equivalent anti-tumor activity. *Cancer Immunology Research, 1*, 99–111. http://dx.doi.org/10.1158/2326-6066.CIR-13-0047.

Dougan, S. K., Ogata, S., Hu, C. C., Grotenbreg, G. M., Guillen, E., Jaenisch, R., et al. (2012). IgG1 + ovalbumin-specific B-cell transnuclear mice show class switch recombination in rare allelically included B cells. *Proceedings of the National Academy of Sciences of the United States of America, 109*(34), 13739–13744. http://dx.doi.org/10.1073/pnas.1210273109.

Duhren-von Minden, M., Ubelhart, R., Schneider, D., Wossning, T., Bach, M. P., Buchner, M., et al. (2012). Chronic lymphocytic leukaemia is driven by antigen-independent cell-autonomous signalling. *Nature, 489*(7415), 309–312. http://dx.doi.org/10.1038/nature11309.

Ehlers, M., Fukuyama, H., McGaha, T. L., Aderem, A., & Ravetch, J. V. (2006). TLR9/MyD88 signaling is required for class switching to pathogenic IgG2a and 2b autoantibodies in SLE. *The Journal of Experimental Medicine, 203*(3), 553–561. http://dx.doi.org/10.1084/jem.20052438.

Engel, P., Zhou, L. J., Ord, D. C., Sato, S., Koller, B., & Tedder, T. F. (1995). Abnormal B lymphocyte development, activation, and differentiation in mice that lack or overexpress the CD19 signal transduction molecule. *Immunity, 3*(1), 39–50.

Erikson, J., Radic, M. Z., Camper, S. A., Hardy, R. R., Carmack, C., & Weigert, M. (1991). Expression of anti-DNA immunoglobulin transgenes in non-autoimmune mice. *Nature, 349*(6307), 331–334. http://dx.doi.org/10.1038/349331a0.

Fais, F., Ghiotto, F., Hashimoto, S., Sellars, B., Valetto, A., Allen, S. L., et al. (1998). Chronic lymphocytic leukemia B cells express restricted sets of mutated and unmutated antigen receptors. *The Journal of Clinical Investigation, 102*(8), 1515–1525. http://dx.doi.org/10.1172/JCI3009.

Flaswinkel, H., & Reth, M. (1994). Dual role of the tyrosine activation motif of the Ig-alpha protein during signal transduction via the B cell antigen receptor. *The EMBO Journal, 13*(1), 83–89.

Fleire, S. J., Goldman, J. P., Carrasco, Y. R., Weber, M., Bray, D., & Batista, F. D. (2006). B cell ligand discrimination through a spreading and contraction response. *Science, 312*(5774), 738–741. http://dx.doi.org/10.1126/science.1123940.

Flores-Borja, F., Kabouridis, P. S., Jury, E. C., Isenberg, D. A., & Mageed, R. A. (2005). Decreased Lyn expression and translocation to lipid raft signaling domains in

B lymphocytes from patients with systemic lupus erythematosus. *Arthritis and Rheumatism*, *52*(12), 3955–3965. http://dx.doi.org/10.1002/art.21416.

Franke, W. W., Schmid, E., Osborn, M., & Weber, K. (1978). Different intermediate-sized filaments distinguished by immunofluorescence microscopy. *Proceedings of the National Academy of Sciences of the United States of America*, *75*(10), 5034–5038.

Fukui, R., Saitoh, S., Kanno, A., Onji, M., Shibata, T., Ito, A., et al. (2011). Unc93B1 restricts systemic lethal inflammation by orchestrating Toll-like receptor 7 and 9 trafficking. *Immunity*, *35*(1), 69–81. http://dx.doi.org/10.1016/j.immuni.2011.05.010.

Fukui, R., Saitoh, S., Matsumoto, F., Kozuka-Hata, H., Oyama, M., Tabeta, K., et al. (2009). Unc93B1 biases Toll-like receptor responses to nucleic acid in dendritic cells toward DNA- but against RNA-sensing. *The Journal of Experimental Medicine*, *206*(6), 1339–1350. http://dx.doi.org/10.1084/jem.20082316.

Gay, D., Saunders, T., Camper, S., & Weigert, M. (1993). Receptor editing: An approach by autoreactive B cells to escape tolerance. *The Journal of Experimental Medicine*, *177*(4), 999–1008.

Georgopoulos, K., Bigby, M., Wang, J. H., Molnar, A., Wu, P., Winandy, S., et al. (1994). The Ikaros gene is required for the development of all lymphoid lineages. *Cell*, *79*(1), 143–156.

Ghiotto, F., Fais, F., Valetto, A., Albesiano, E., Hashimoto, S., Dono, M., et al. (2004). Remarkably similar antigen receptors among a subset of patients with chronic lymphocytic leukemia. *The Journal of Clinical Investigation*, *113*(7), 1008–1016. http://dx.doi.org/10.1172/JCI19399.

Goodnow, C. C., Crosbie, J., Adelstein, S., Lavoie, T. B., Smith-Gill, S. J., Brink, R. A., et al. (1988). Altered immunoglobulin expression and functional silencing of self-reactive B lymphocytes in transgenic mice. *Nature*, *334*(6184), 676–682. http://dx.doi.org/10.1038/334676a0.

Goodnow, C. C., Vinuesa, C. G., Randall, K. L., Mackay, F., & Brink, R. (2010). Control systems and decision making for antibody production. *Nature Immunology*, *11*(8), 681–688. http://dx.doi.org/10.1038/ni.1900.

Grimaldi, C. M., Hicks, R., & Diamond, B. (2005). B cell selection and susceptibility to autoimmunity. *Journal of Immunology*, *174*(4), 1775–1781 (Review).

Halcomb, K. E., Musuka, S., Gutierrez, T., Wright, H. L., & Satterthwaite, A. B. (2008). Btk regulates localization, in vivo activation, and class switching of anti-DNA B cells. *Molecular Immunology*, *46*(2), 233–241. http://dx.doi.org/10.1016/j.molimm.2008.08.278.

Hamblin, T. J., Davis, Z., Gardiner, A., Oscier, D. G., & Stevenson, F. K. (1999). Unmutated Ig V(H) genes are associated with a more aggressive form of chronic lymphocytic leukemia. *Blood*, *94*(6), 1848–1854.

Hanahan, D., & Weinberg, R. A. (2011). Hallmarks of cancer: The next generation. *Cell*, *144*(5), 646–674. http://dx.doi.org/10.1016/j.cell.2011.02.013.

Hannum, L. G., Ni, D., Haberman, A. M., Weigert, M. G., & Shlomchik, M. J. (1996). A disease-related rheumatoid factor autoantibody is not tolerized in a normal mouse: Implications for the origins of autoantibodies in autoimmune disease. *The Journal of Experimental Medicine*, *184*(4), 1269–1278.

Hantschel, O., Rix, U., Schmidt, U., Burckstummer, T., Kneidinger, M., Schutze, G., et al. (2007). The Btk tyrosine kinase is a major target of the Bcr-Abl inhibitor dasatinib. *Proceedings of the National Academy of Sciences of the United States of America*, *104*(33), 13283–13288. http://dx.doi.org/10.1073/pnas.0702654104.

Hara, H., Wada, T., Bakal, C., Kozieradzki, I., Suzuki, S., Suzuki, N., et al. (2003). The MAGUK family protein CARD11 is essential for lymphocyte activation. *Immunity*, *18*(6), 763–775.

Hardy, R. R., & Hayakawa, K. (2001). B cell development pathways. *Annual Review of Immunology*, *19*, 595–621. http://dx.doi.org/10.1146/annurev.immunol.19.1.595.

Hartley, S. B., Crosbie, J., Brink, R., Kantor, A. B., Basten, A., & Goodnow, C. C. (1991). Elimination from peripheral lymphoid tissues of self-reactive B lymphocytes recognizing membrane-bound antigens. *Nature, 353*(6346), 765–769. http://dx.doi.org/10.1038/353765a0.

Hermine, O., Lefrere, F., Bronowicki, J. P., Mariette, X., Jondeau, K., Eclache-Saudreau, V., et al. (2002). Regression of splenic lymphoma with villous lymphocytes after treatment of hepatitis C virus infection. *The New England Journal of Medicine, 347*(2), 89–94. http://dx.doi.org/10.1056/NEJMoa013376.

Herve, M., Xu, K., Ng, Y. S., Wardemann, H., Albesiano, E., Messmer, B. T., et al. (2005). Unmutated and mutated chronic lymphocytic leukemias derive from self-reactive B cell precursors despite expressing different antibody reactivity. *The Journal of Clinical Investigation, 115*(6), 1636–1643. http://dx.doi.org/10.1172/JCI24387.

Herzog, S., & Jumaa, H. (2012). Self-recognition and clonal selection: Autoreactivity drives the generation of B cells. *Current Opinion in Immunology, 24*(2), 166–172. http://dx.doi.org/10.1016/j.coi.2012.02.004.

Hibbs, M. L., Harder, K. W., Armes, J., Kountouri, N., Quilici, C., Casagranda, F., et al. (2002). Sustained activation of Lyn tyrosine kinase in vivo leads to autoimmunity. *The Journal of Experimental Medicine, 196*(12), 1593–1604.

Hibbs, M. L., Tarlinton, D. M., Armes, J., Grail, D., Hodgson, G., Maglitto, R., et al. (1995). Multiple defects in the immune system of Lyn-deficient mice, culminating in autoimmune disease. *Cell, 83*(2), 301–311.

Hochedlinger, K., & Jaenisch, R. (2002). Monoclonal mice generated by nuclear transfer from mature B and T donor cells. *Nature, 415*(6875), 1035–1038. http://dx.doi.org/10.1038/nature718.

Hombach, J., Tsubata, T., Leclercq, L., Stappert, H., & Reth, M. (1990). Molecular components of the B-cell antigen receptor complex of the IgM class. *Nature, 343*(6260), 760–762. http://dx.doi.org/10.1038/343760a0.

Honigberg, L. A., Smith, A. M., Sirisawad, M., Verner, E., Loury, D., Chang, B., et al. (2010). The Bruton tyrosine kinase inhibitor PCI-32765 blocks B-cell activation and is efficacious in models of autoimmune disease and B-cell malignancy. *Proceedings of the National Academy of Sciences of the United States of America, 107*(29), 13075–13080. http://dx.doi.org/10.1073/pnas.1004594107.

Hoogeboom, R., van Kessel, K. P., Hochstenbach, F., Wormhoudt, T. A., Reinten, R. J., Wagner, K., et al. (2013). A mutated B cell chronic lymphocytic leukemia subset that recognizes and responds to fungi. *The Journal of Experimental Medicine, 210*(1), 59–70. http://dx.doi.org/10.1084/jem.20121801.

Joshi, I., Yoshida, T., Jena, N., Qi, X., Zhang, J., Van Etten, R. A., et al. (2014). Loss of Ikaros DNA-binding function confers integrin-dependent survival on pre-B cells and progression to acute lymphoblastic leukemia. *Nature Immunology, 15*, 294–304. http://dx.doi.org/10.1038/ni.2821.

Jun, J. E., Wilson, L. E., Vinuesa, C. G., Lesage, S., Blery, M., Miosge, L. A., et al. (2003). Identifying the MAGUK protein Carma-1 as a central regulator of humoral immune responses and atopy by genome-wide mouse mutagenesis. *Immunity, 18*(6), 751–762.

Kim, Y. M., Brinkmann, M. M., Paquet, M. E., & Ploegh, H. L. (2008). UNC93B1 delivers nucleotide-sensing toll-like receptors to endolysosomes. *Nature, 452*(7184), 234–238. http://dx.doi.org/10.1038/nature06726.

Kim, Y. M., Pan, J. Y., Korbel, G. A., Peperzak, V., Boes, M., & Ploegh, H. L. (2006). Monovalent ligation of the B cell receptor induces receptor activation but fails to promote antigen presentation. *Proceedings of the National Academy of Sciences of the United States of America, 103*(9), 3327–3332. http://dx.doi.org/10.1073/pnas.0511315103.

Kirak, O., Frickel, E. M., Grotenbreg, G. M., Suh, H., Jaenisch, R., & Ploegh, H. L. (2010a). Transnuclear mice with pre-defined T cell receptor specificities against Toxoplasma gondii obtained via SCNT. *Journal of Visualized Experiments*. (43). http://dx.doi.org/10.3791/2168.

Kirak, O., Frickel, E. M., Grotenbreg, G. M., Suh, H., Jaenisch, R., & Ploegh, H. L. (2010b). Transnuclear mice with predefined T cell receptor specificities against Toxoplasma gondii obtained via SCNT. *Science*, *328*(5975), 243–248. http://dx.doi.org/10.1126/science.1178590.

Kishigami, S., Wakayama, S., Thuan, N. V., Ohta, H., Mizutani, E., Hikichi, T., et al. (2006). Production of cloned mice by somatic cell nuclear transfer. *Nature Protocols*, *1*(1), 125–138. http://dx.doi.org/10.1038/nprot.2006.21.

Klapper, W., Szczepanowski, M., Burkhardt, B., Berger, H., Rosolowski, M., Bentink, S., et al. (2008). Molecular profiling of pediatric mature B-cell lymphoma treated in population-based prospective clinical trials. *Blood*, *112*(4), 1374–1381. http://dx.doi.org/10.1182/blood-2008-01-136465.

Klein, U., Klein, G., Ehlin-Henriksson, B., Rajewsky, K., & Kuppers, R. (1995). Burkitt's lymphoma is a malignancy of mature B cells expressing somatically mutated V region genes. *Molecular Medicine*, *1*(5), 495–505.

Kridel, R., Sehn, L. H., & Gascoyne, R. D. (2012). Pathogenesis of follicular lymphoma. *The Journal of Clinical Investigation*, *122*(10), 3424–3431. http://dx.doi.org/10.1172/JCI63186.

Krieg, A. M., Yi, A. K., Matson, S., Waldschmidt, T. J., Bishop, G. A., Teasdale, R., et al. (1995). CpG motifs in bacterial DNA trigger direct B-cell activation. *Nature*, *374*(6522), 546–549. http://dx.doi.org/10.1038/374546a0.

Krysov, S., Potter, K. N., Mockridge, C. I., Coelho, V., Wheatley, I., Packham, G., et al. (2010). Surface IgM of CLL cells displays unusual glycans indicative of engagement of antigen in vivo. *Blood*, *115*(21), 4198–4205. http://dx.doi.org/10.1182/blood-2009-12-254847.

Kuppers, R. (2005). Mechanisms of B-cell lymphoma pathogenesis. *Nature Reviews. Cancer*, *5*(4), 251–262. http://dx.doi.org/10.1038/nrc1589.

Kurosaki, T., Shinohara, H., & Baba, Y. (2010). B cell signaling and fate decision. *Annual Review of Immunology*, *28*, 21–55. http://dx.doi.org/10.1146/annurev.immunol.021908.132541.

Lam, K. P., Kuhn, R., & Rajewsky, K. (1997). In vivo ablation of surface immunoglobulin on mature B cells by inducible gene targeting results in rapid cell death. *Cell*, *90*(6), 1073–1083.

Lam, L. T., Wright, G., Davis, R. E., Lenz, G., Farinha, P., Dang, L., et al. (2008). Cooperative signaling through the signal transducer and activator of transcription 3 and nuclear factor-{kappa}B pathways in subtypes of diffuse large B-cell lymphoma. *Blood*, *111*(7), 3701–3713. http://dx.doi.org/10.1182/blood-2007-09-111948.

Lanemo Myhrinder, A., Hellqvist, E., Sidorova, E., Soderberg, A., Baxendale, H., Dahle, C., et al. (2008). A new perspective: Molecular motifs on oxidized LDL, apoptotic cells, and bacteria are targets for chronic lymphocytic leukemia antibodies. *Blood*, *111*(7), 3838–3848. http://dx.doi.org/10.1182/blood-2007-11-125450.

Lau, C. M., Broughton, C., Tabor, A. S., Akira, S., Flavell, R. A., Mamula, M. J., et al. (2005). RNA-associated autoantigens activate B cells by combined B cell antigen receptor/Toll-like receptor 7 engagement. *The Journal of Experimental Medicine*, *202*(9), 1171–1177. http://dx.doi.org/10.1084/jem.20050630.

Leadbetter, E. A., Rifkin, I. R., Hohlbaum, A. M., Beaudette, B. C., Shlomchik, M. J., & Marshak-Rothstein, A. (2002). Chromatin-IgG complexes activate B cells by dual engagement of IgM and Toll-like receptors. *Nature*, *416*(6881), 603–607.

Lee, C. C., Avalos, A. M., & Ploegh, H. L. (2012). Accessory molecules for Toll-like receptors and their function. *Nature Reviews. Immunology*, *12*(3), 168–179. http://dx.doi.org/10.1038/nri3151.

Lee, B. L., Moon, J. E., Shu, J. H., Yuan, L., Newman, Z. R., Schekman, R., et al. (2013). UNC93B1 mediates differential trafficking of endosomal TLRs. *eLife*, *2*, e00291. http://dx.doi.org/10.7554/eLife.00291.

Lenz, G., Davis, R. E., Ngo, V. N., Lam, L., George, T. C., Wright, G. W., et al. (2008). Oncogenic CARD11 mutations in human diffuse large B cell lymphoma. *Science*, *319*(5870), 1676–1679. http://dx.doi.org/10.1126/science.1153629.

Lenz, G., Wright, G., Dave, S. S., Xiao, W., Powell, J., Zhao, H., et al. (2008). Stromal gene signatures in large-B-cell lymphomas. *The New England Journal of Medicine*, *359*(22), 2313–2323. http://dx.doi.org/10.1056/NEJMoa0802885.

Lenze, D., Berg, E., Volkmer-Engert, R., Weiser, A. A., Greiner, A., Knorr-Wittmann, C., et al. (2006). Influence of antigen on the development of MALT lymphoma. *Blood*, *107*(3), 1141–1148. http://dx.doi.org/10.1182/blood-2005-04-1722.

Lu, R., Vidal, G. S., Kelly, J. A., Delgado-Vega, A. M., Howard, X. K., Macwana, S. R., et al. (2009). Genetic associations of LYN with systemic lupus erythematosus. *Genes and Immunity*, *10*(5), 397–403. http://dx.doi.org/10.1038/gene.2009.19.

Ma, H., Qazi, S., Ozer, Z., Gaynon, P., Reaman, G. H., & Uckun, F. M. (2012). CD22 exon 12 deletion is a characteristic genetic defect of therapy-refractory clones in paediatric acute lymphoblastic leukaemia. *British Journal of Haematology*, *156*(1), 89–98. http://dx.doi.org/10.1111/j.1365-2141.2011.08901.x.

Mackay, M., Stanevsky, A., Wang, T., Aranow, C., Li, M., Koenig, S., et al. (2006). Selective dysregulation of the FcgammaIIB receptor on memory B cells in SLE. *The Journal of Experimental Medicine*, *203*(9), 2157–2164. http://dx.doi.org/10.1084/jem.20051503.

Maloney, D. G. (2012). Anti-CD20 antibody therapy for B-cell lymphomas. *The New England Journal of Medicine*, *366*(21), 2008–2016. http://dx.doi.org/10.1056/NEJMct1114348.

Mandik-Nayak, L., Seo, S. J., Sokol, C., Potts, K. M., Bui, A., & Erikson, J. (1999). MRL-lpr/lpr mice exhibit a defect in maintaining developmental arrest and follicular exclusion of anti-double-stranded DNA B cells. *The Journal of Experimental Medicine*, *189*(11), 1799–1814.

Mann, D. L., DeSantis, P., Mark, G., Pfeifer, A., Newman, M., Gibbs, N., et al. (1987). HTLV-I-associated B-cell CLL: Indirect role for retrovirus in leukemogenesis. *Science*, *236*(4805), 1103–1106.

Marcucci, F., & Mele, A. (2011). Hepatitis viruses and non-Hodgkin lymphoma: Epidemiology, mechanisms of tumorigenesis, and therapeutic opportunities. *Blood*, *117*(6), 1792–1798. http://dx.doi.org/10.1182/blood-2010-06-275818.

Marshak-Rothstein, A. (2006). Toll-like receptors in systemic autoimmune disease. *Nature Reviews. Immunology*, *6*(11), 823–835. http://dx.doi.org/10.1038/nri1957.

Martin, F., Oliver, A. M., & Kearney, J. F. (2001). Marginal zone and B1 B cells unite in the early response against T-independent blood-borne particulate antigens. *Immunity*, *14*(5), 617–629.

Mattila, P. K., Feest, C., Depoil, D., Treanor, B., Montaner, B., Otipoby, K. L., et al. (2013). The actin and tetraspanin networks organize receptor nanoclusters to regulate B cell receptor-mediated signaling. *Immunity*, *38*(3), 461–474. http://dx.doi.org/10.1016/j.immuni.2012.11.019.

McGaha, T. L., Sorrentino, B., & Ravetch, J. V. (2005). Restoration of tolerance in lupus by targeted inhibitory receptor expression. *Science*, *307*(5709), 590–593. http://dx.doi.org/10.1126/science.1105160.

Meffre, E., & Wardemann, H. (2008). B-cell tolerance checkpoints in health and autoimmunity. *Current Opinion in Immunology*, *20*(6), 632–638. http://dx.doi.org/10.1016/j.coi.2008.09.001.

Merrell, K. T., Benschop, R. J., Gauld, S. B., Aviszus, K., Decote-Ricardo, D., Wysocki, L. J., et al. (2006). Identification of anergic B cells within a wild-type repertoire. *Immunity*, *25*(6), 953–962. http://dx.doi.org/10.1016/j.immuni.2006.10.017.

Messmer, B. T., Albesiano, E., Efremov, D. G., Ghiotto, F., Allen, S. L., Kolitz, J., et al. (2004). Multiple distinct sets of stereotyped antigen receptors indicate a role for antigen in promoting chronic lymphocytic leukemia. *The Journal of Experimental Medicine*, *200*(4), 519–525. http://dx.doi.org/10.1084/jem.20040544.

Miles, R. R., Raphael, M., McCarthy, K., Wotherspoon, A., Lones, M. A., Terrier-Lacombe, M. J., et al. (2008). Pediatric diffuse large B-cell lymphoma demonstrates a high proliferation index, frequent c-Myc protein expression, and a high incidence of germinal center subtype: Report of the French-American-British (FAB) international study group. *Pediatric Blood & Cancer*, *51*(3), 369–374. http://dx.doi.org/10.1002/pbc.21619.

Minguet, S., Dopfer, E. P., & Schamel, W. W. (2010). Low-valency, but not monovalent, antigens trigger the B-cell antigen receptor (BCR). *International Immunology*, *22*(3), 205–212. http://dx.doi.org/10.1093/intimm/dxp129.

Mockridge, C. I., Potter, K. N., Wheatley, I., Neville, L. A., Packham, G., & Stevenson, F. K. (2007). Reversible anergy of sIgM-mediated signaling in the two subsets of CLL defined by VH-gene mutational status. *Blood*, *109*(10), 4424–4431. http://dx.doi.org/10.1182/blood-2006-11-056648.

Monroe, J. G. (2006). ITAM-mediated tonic signalling through pre-BCR and BCR complexes. *Nature Reviews. Immunology*, *6*(4), 283–294. http://dx.doi.org/10.1038/nri1808.

Mullighan, C. G., Su, X., Zhang, J., Radtke, I., Phillips, L. A., Miller, C. B., et al. (2009). Deletion of IKZF1 and prognosis in acute lymphoblastic leukemia. *The New England Journal of Medicine*, *360*(5), 470–480. http://dx.doi.org/10.1056/NEJMoa0808253.

Muzio, M., Apollonio, B., Scielzo, C., Frenquelli, M., Vandoni, I., Boussiotis, V., et al. (2008). Constitutive activation of distinct BCR-signaling pathways in a subset of CLL patients: A molecular signature of anergy. *Blood*, *112*(1), 188–195. http://dx.doi.org/10.1182/blood-2007-09-111344.

Nakayama, J., Yamamoto, M., Hayashi, K., Satoh, H., Bundo, K., Kubo, M., et al. (2009). BLNK suppresses pre-B-cell leukemogenesis through inhibition of JAK3. *Blood*, *113*(7), 1483–1492. http://dx.doi.org/10.1182/blood-2008-07-166355.

Nath, S. K., Han, S., Kim-Howard, X., Kelly, J. A., Viswanathan, P., Gilkeson, G. S., et al. (2008). A nonsynonymous functional variant in integrin-alpha(M) (encoded by ITGAM) is associated with systemic lupus erythematosus. *Nature Genetics*, *40*(2), 152–154. http://dx.doi.org/10.1038/ng.71.

Nemazee, D. A., & Burki, K. (1989). Clonal deletion of B lymphocytes in a transgenic mouse bearing anti-MHC class I antibody genes. *Nature*, *337*(6207), 562–566. http://dx.doi.org/10.1038/337562a0.

Ngo, V. N., Davis, R. E., Lamy, L., Yu, X., Zhao, H., Lenz, G., et al. (2006). A loss-of-function RNA interference screen for molecular targets in cancer. *Nature*, *441*(7089), 106–110. http://dx.doi.org/10.1038/nature04687.

Ngo, V. N., Young, R. M., Schmitz, R., Jhavar, S., Xiao, W., Lim, K. H., et al. (2011). Oncogenically active MYD88 mutations in human lymphoma. *Nature*, *470*(7332), 115–119. http://dx.doi.org/10.1038/nature09671.

Nickerson, K. M., Christensen, S. R., Shupe, J., Kashgarian, M., Kim, D., Elkon, K., et al. (2010). TLR9 regulates TLR7- and MyD88-dependent autoantibody production and disease in a murine model of lupus. *Journal of Immunology*, *184*(4), 1840–1848. http://dx.doi.org/10.4049/jimmunol.0902592.

Nimmerjahn, F., Bruhns, P., Horiuchi, K., & Ravetch, J. V. (2005). FcgammaRIV: A novel FcR with distinct IgG subclass specificity. *Immunity*, *23*(1), 41–51. http://dx.doi.org/10.1016/j.immuni.2005.05.010.

Nishizumi, H., Taniuchi, I., Yamanashi, Y., Kitamura, D., Ilic, D., Mori, S., et al. (1995). Impaired proliferation of peripheral B cells and indication of autoimmune disease in lyn-deficient mice. *Immunity*, *3*(5), 549–560.

Nitschke, L., Carsetti, R., Ocker, B., Kohler, G., & Lamers, M. C. (1997). CD22 is a negative regulator of B-cell receptor signalling. *Current Biology*, 7(2), 133–143.
Nundel, K., Busto, P., Debatis, M., & Marshak-Rothstein, A. (2013). The role of Bruton's tyrosine kinase in the development and BCR/TLR-dependent activation of AM14 rheumatoid factor B cells. *Journal of Leukocyte Biology*, *94*(5), 865–875. http://dx.doi.org/10.1189/jlb.0313126.
Nussenzweig, A., & Nussenzweig, M. C. (2010). Origin of chromosomal translocations in lymphoid cancer. *Cell*, *141*(1), 27–38. http://dx.doi.org/10.1016/j.cell.2010.03.016.
O'Keefe, T. L., Williams, G. T., Batista, F. D., & Neuberger, M. S. (1999). Deficiency in CD22, a B cell-specific inhibitory receptor, is sufficient to predispose to development of high affinity autoantibodies. *The Journal of Experimental Medicine*, *189*(8), 1307–1313.
O'Keefe, T. L., Williams, G. T., Davies, S. L., & Neuberger, M. S. (1996). Hyperresponsive B cells in CD22-deficient mice. *Science*, *274*(5288), 798–801.
Oliver, A. M., Martin, F., Gartland, G. L., Carter, R. H., & Kearney, J. F. (1997). Marginal zone B cells exhibit unique activation, proliferative and immunoglobulin secretory responses. *European Journal of Immunology*, *27*(9), 2366–2374. http://dx.doi.org/10.1002/eji.1830270935.
Oschlies, I., Klapper, W., Zimmermann, M., Krams, M., Wacker, H. H., Burkhardt, B., et al. (2006). Diffuse large B-cell lymphoma in pediatric patients belongs predominantly to the germinal-center type B-cell lymphomas: A clinicopathologic analysis of cases included in the German BFM (Berlin-Frankfurt-Munster) Multicenter Trial. *Blood*, *107*(10), 4047–4052. http://dx.doi.org/10.1182/blood-2005-10-4213.
Pasqualucci, L., Neumeister, P., Goossens, T., Nanjangud, G., Chaganti, R. S., Kuppers, R., et al. (2001). Hypermutation of multiple proto-oncogenes in B-cell diffuse large-cell lymphomas. *Nature*, *412*(6844), 341–346. http://dx.doi.org/10.1038/35085588.
Paz, Z., & Tsokos, G. C. (2013). New therapeutics in systemic lupus erythematosus. *Current Opinion in Rheumatology*, *25*(3), 297–303. http://dx.doi.org/10.1097/BOR.0b013e32835fd682.
Pillai, S., Cariappa, A., & Moran, S. T. (2005). Marginal zone B cells. *Annual Review of Immunology*, *23*, 161–196. http://dx.doi.org/10.1146/annurev.immunol.23.021704.115728.
Quinn, E. R., Chan, C. H., Hadlock, K. G., Foung, S. K., Flint, M., & Levy, S. (2001). The B-cell receptor of a hepatitis C virus (HCV)-associated non-Hodgkin lymphoma binds the viral E2 envelope protein, implicating HCV in lymphomagenesis. *Blood*, *98*(13), 3745–3749.
Radcliffe, C. M., Arnold, J. N., Suter, D. M., Wormald, M. R., Harvey, D. J., Royle, L., et al. (2007). Human follicular lymphoma cells contain oligomannose glycans in the antigen-binding site of the B-cell receptor. *The Journal of Biological Chemistry*, *282*(10), 7405–7415. http://dx.doi.org/10.1074/jbc.M602690200.
Radic, M. Z., Erikson, J., Litwin, S., & Weigert, M. (1993). B lymphocytes may escape tolerance by revising their antigen receptors. *The Journal of Experimental Medicine*, *177*(4), 1165–1173.
Rahman, A., & Isenberg, D. A. (2008). Systemic lupus erythematosus. *The New England Journal of Medicine*, *358*(9), 929–939. http://dx.doi.org/10.1056/NEJMra071297.
Rankin, A. L., Seth, N., Keegan, S., Andreyeva, T., Cook, T. A., Edmonds, J., et al. (2013). Selective inhibition of BTK prevents murine lupus and antibody-mediated glomerulonephritis. *Journal of Immunology*, *191*(9), 4540–4550. http://dx.doi.org/10.4049/jimmunol.1301553.
Rickert, R. C., Rajewsky, K., & Roes, J. (1995). Impairment of T-cell-dependent B-cell responses and B-1 cell development in CD19-deficient mice. *Nature*, *376*(6538), 352–355. http://dx.doi.org/10.1038/376352a0.
Rosenwald, A., Wright, G., Chan, W. C., Connors, J. M., Campo, E., Fisher, R. I., et al. (2002). The use of molecular profiling to predict survival after chemotherapy for diffuse

large-B-cell lymphoma. *The New England Journal of Medicine, 346*(25), 1937–1947. http://dx.doi.org/10.1056/NEJMoa012914.
Rullo, O. J., & Tsao, B. P. (2013). Recent insights into the genetic basis of systemic lupus erythematosus. *Annals of the Rheumatic Diseases, 72*(Suppl 2), ii56–ii61. http://dx.doi.org/10.1136/annrheumdis-2012-202351.
Russell, S. M., Tayebi, N., Nakajima, H., Riedy, M. C., Roberts, J. L., Aman, M. J., et al. (1995). Mutation of Jak3 in a patient with SCID: Essential role of Jak3 in lymphoid development. *Science, 270*(5237), 797–800.
Sachen, K. L., Strohman, M. J., Singletary, J., Alizadeh, A. A., Kattah, N. H., Lossos, C., et al. (2012). Self-antigen recognition by follicular lymphoma B-cell receptors. *Blood, 120*(20), 4182–4190. http://dx.doi.org/10.1182/blood-2012-05-427534.
Santulli-Marotto, S., Qian, Y., Ferguson, S., & Clarke, S. H. (2001). Anti-Sm B cell differentiation in Ig transgenic MRL/Mp-lpr/lpr mice: Altered differentiation and an accelerated response. *Journal of Immunology, 166*(8), 5292–5299.
Santulli-Marotto, S., Retter, M. W., Gee, R., Mamula, M. J., & Clarke, S. H. (1998). Autoreactive B cell regulation: Peripheral induction of developmental arrest by lupus-associated autoantigens. *Immunity, 8*(2), 209–219.
Sato, S., Ono, N., Steeber, D. A., Pisetsky, D. S., & Tedder, T. F. (1996). CD19 regulates B lymphocyte signaling thresholds critical for the development of B-1 lineage cells and autoimmunity. *Journal of Immunology, 157*(10), 4371–4378.
Schmitz, R., Young, R. M., Ceribelli, M., Jhavar, S., Xiao, W., Zhang, M., et al. (2012). Burkitt lymphoma pathogenesis and therapeutic targets from structural and functional genomics. *Nature, 490*(7418), 116–120. http://dx.doi.org/10.1038/nature11378.
Schwering, I., Brauninger, A., Klein, U., Jungnickel, B., Tinguely, M., Diehl, V., et al. (2003). Loss of the B-lineage-specific gene expression program in Hodgkin and Reed-Sternberg cells of Hodgkin lymphoma. *Blood, 101*(4), 1505–1512. http://dx.doi.org/10.1182/blood-2002-03-0839.
Schwickert, T. A., Tagoh, H., Gultekin, S., Dakic, A., Axelsson, E., Minnich, M., et al. (2014). Stage-specific control of early B cell development by the transcription factor Ikaros. *Nature Immunology, 15*, 283–293. http://dx.doi.org/10.1038/ni.2828.
Sehrawat, S., Kirak, O., Koenig, P. A., Isaacson, M. K., Marques, S., Bozkurt, G., et al. (2012). CD8(+) T cells from mice transnuclear for a TCR that recognizes a single H-2K(b)-restricted MHV68 epitope derived from gB-ORF8 help control infection. *Cell Reports, 1*(5), 461–471. http://dx.doi.org/10.1016/j.celrep.2012.03.009.
Seldin, M. F., Reeves, J. P., Scribner, C. L., Roths, J. B., Davidson, W. F., Morse, H. C., 3rd, et al. (1987). Effect of xid on autoimmune C3H-gld/gld mice. *Cellular Immunology, 107*(1), 249–255.
Shaffer, A. L., & Schlissel, M. S. (1997). A truncated heavy chain protein relieves the requirement for surrogate light chains in early B cell development. *The Journal of Immunology, 159*(3), 1265–1275.
Shaffer, A. L., Yu, X., He, Y., Boldrick, J., Chan, E. P., & Staudt, L. M. (2000). BCL-6 represses genes that function in lymphocyte differentiation, inflammation, and cell cycle control. *Immunity, 13*(2), 199–212.
Shlomchik, M. J. (2008). Sites and stages of autoreactive B cell activation and regulation. *Immunity, 28*(1), 18–28. http://dx.doi.org/10.1016/j.immuni.2007.12.004.
Shlomchik, M. J., Zharhary, D., Saunders, T., Camper, S. A., & Weigert, M. G. (1993). A rheumatoid factor transgenic mouse model of autoantibody regulation. *International Immunology, 5*(10), 1329–1341.
Smedby, K. E., Hjalgrim, H., Askling, J., Chang, E. T., Gregersen, H., Porwit-MacDonald, A., et al. (2006). Autoimmune and chronic inflammatory disorders and risk of non-Hodgkin lymphoma by subtype. *Journal of the National Cancer Institute, 98*(1), 51–60. http://dx.doi.org/10.1093/jnci/djj004.

Staudt, L. M. (2007). A closer look at follicular lymphoma. *The New England Journal of Medicine, 356*(7), 741–742. http://dx.doi.org/10.1056/NEJMcibr067155.
Steinberg, B. J., Smathers, P. A., Frederiksen, K., & Steinberg, A. D. (1982). Ability of the xid gene to prevent autoimmunity in (NZB X NZW)F1 mice during the course of their natural history, after polyclonal stimulation, or following immunization with DNA. *The Journal of Clinical Investigation, 70*(3), 587–597.
Sthoeger, Z. M., Wakai, M., Tse, D. B., Vinciguerra, V. P., Allen, S. L., Budman, D. R., et al. (1989). Production of autoantibodies by CD5-expressing B lymphocytes from patients with chronic lymphocytic leukemia. *The Journal of Experimental Medicine, 169*(1), 255–268.
Stoddart, A., Dykstra, M. L., Brown, B. K., Song, W., Pierce, S. K., & Brodsky, F. M. (2002). Lipid rafts unite signaling cascades with clathrin to regulate BCR internalization. *Immunity, 17*(4), 451–462.
Swerdlow, S., Campo, E., Harris, N. L., Jaffe, E. S., Pileri, S., Stein, H., et al. (2008). *WHO classification of tumours of haematopoietic and lymphoid tissues* (4th ed.). Lyon: International Agency for Research on Cancer (IARC).
Tedder, T. F., Inaoki, M., & Sato, S. (1997). The CD19-CD21 complex regulates signal transduction thresholds governing humoral immunity and autoimmunity. *Immunity, 6*(2), 107–118.
Tiegs, S. L., Russell, D. M., & Nemazee, D. (1993). Receptor editing in self-reactive bone marrow B cells. *The Journal of Experimental Medicine, 177*(4), 1009–1020.
Tiller, T., Tsuiji, M., Yurasov, S., Velinzon, K., Nussenzweig, M. C., & Wardemann, H. (2007). Autoreactivity in human IgG+ memory B cells. *Immunity, 26*(2), 205–213. http://dx.doi.org/10.1016/j.immuni.2007.01.009.
Tobin, G., Thunberg, U., Johnson, A., Eriksson, I., Soderberg, O., Karlsson, K., et al. (2003). Chronic lymphocytic leukemias utilizing the VH3-21 gene display highly restricted Vlambda2-14 gene use and homologous CDR3s: Implicating recognition of a common antigen epitope. *Blood, 101*(12), 4952–4957. http://dx.doi.org/10.1182/blood-2002-11-3485.
Tolar, P., Hanna, J., Krueger, P. D., & Pierce, S. K. (2009). The constant region of the membrane immunoglobulin mediates B cell-receptor clustering and signaling in response to membrane antigens. *Immunity, 30*(1), 44–55. http://dx.doi.org/10.1016/j.immuni.2008.11.007.
Tolar, P., Sohn, H. W., & Pierce, S. K. (2005). The initiation of antigen-induced B cell antigen receptor signaling viewed in living cells by fluorescence resonance energy transfer. *Nature Immunology, 6*(11), 1168–1176. http://dx.doi.org/10.1038/ni1262.
Treanor, B., Depoil, D., Gonzalez-Granja, A., Barral, P., Weber, M., Dushek, O., et al. (2010). The membrane skeleton controls diffusion dynamics and signaling through the B cell receptor. *Immunity, 32*(2), 187–199. http://dx.doi.org/10.1016/j.immuni.2009.12.005.
Tsujimoto, Y., Cossman, J., Jaffe, E., & Croce, C. M. (1985). Involvement of the bcl-2 gene in human follicular lymphoma. *Science, 228*(4706), 1440–1443.
Tsujimoto, Y., Gorham, J., Cossman, J., Jaffe, E., & Croce, C. M. (1985). The t(14;18) chromosome translocations involved in B-cell neoplasms result from mistakes in VDJ joining. *Science, 229*(4720), 1390–1393.
Tudor, D., Yu, H., Maupetit, J., Drillet, A. S., Bouceba, T., Schwartz-Cornil, I., et al. (2012). Isotype modulates epitope specificity, affinity, and antiviral activities of anti-HIV-1 human broadly neutralizing 2F5 antibody. *Proceedings of the National Academy of Sciences of the United States of America, 109*(31), 12680–12685. http://dx.doi.org/10.1073/pnas.1200024109.
Uccellini, M. B., Busconi, L., Green, N. M., Busto, P., Christensen, S. R., Shlomchik, M. J., et al. (2008). Autoreactive B cells discriminate CpG-rich and CpG-poor DNA and this

response is modulated by IFN-alpha. *The Journal of Immunology*, *181*(9), 5875–5884 doi: 181/9/5875.

Uckun, F. M., Goodman, P., Ma, H., Dibirdik, I., & Qazi, S. (2010). CD22 EXON 12 deletion as a pathogenic mechanism of human B-precursor leukemia. *Proceedings of the National Academy of Sciences of the United States of America*, *107*(39), 16852–16857. http://dx.doi.org/10.1073/pnas.1007896107.

Vaandrager, J. W., Schuuring, E., Kluin-Nelemans, H. C., Dyer, M. J., Raap, A. K., & Kluin, P. M. (1998). DNA fiber fluorescence in situ hybridization analysis of immunoglobulin class switching in B-cell neoplasia: Aberrant CH gene rearrangements in follicle center-cell lymphoma. *Blood*, *92*(8), 2871–2878.

van Zelm, M. C., Reisli, I., van der Burg, M., Castano, D., van Noesel, C. J., van Tol, M. J., et al. (2006). An antibody-deficiency syndrome due to mutations in the CD19 gene. *The New England Journal of Medicine*, *354*(18), 1901–1912. http://dx.doi.org/10.1056/NEJMoa051568.

Victora, G. D., & Nussenzweig, M. C. (2012). Germinal centers. *Annual Review of Immunology*, *30*, 429–457. http://dx.doi.org/10.1146/annurev-immunol-020711-075032.

Viglianti, G. A., Lau, C. M., Hanley, T. M., Miko, B. A., Shlomchik, M. J., & Marshak-Rothstein, A. (2003). Activation of autoreactive B cells by CpG dsDNA. *Immunity*, *19*(6), 837–847.

Wang, J. Q., Jeelall, Y. S., Beutler, B., Horikawa, K., & Goodnow, C. C. (2014). Consequences of the recurrent MYD88L265P somatic mutation for B cell tolerance. *The Journal of Experimental Medicine*, *211*, 413–426. http://dx.doi.org/10.1084/jem.20131424.

Wang, H., & Shlomchik, M. J. (1997). High affinity rheumatoid factor transgenic B cells are eliminated in normal mice. *Journal of Immunology*, *159*(3), 1125–1134.

Wang, H., & Shlomchik, M. J. (1999). Autoantigen-specific B cell activation in Fas-deficient rheumatoid factor immunoglobulin transgenic mice. *The Journal of Experimental Medicine*, *190*(5), 639–649.

Ward, M. M., Pyun, E., & Studenski, S. (1995). Causes of death in systemic lupus erythematosus. Long-term followup of an inception cohort. *Arthritis and Rheumatism*, *38*(10), 1492–1499.

Wardemann, H., Yurasov, S., Schaefer, A., Young, J. W., Meffre, E., & Nussenzweig, M. C. (2003). Predominant autoantibody production by early human B cell precursors. *Science*, *301*(5638), 1374–1377. http://dx.doi.org/10.1126/science.1086907.

Willcocks, L. C., Carr, E. J., Niederer, H. A., Rayner, T. F., Williams, T. N., Yang, W., et al. (2010). A defunctioning polymorphism in FCGR2B is associated with protection against malaria but susceptibility to systemic lupus erythematosus. *Proceedings of the National Academy of Sciences of the United States of America*, *107*(17), 7881–7885. http://dx.doi.org/10.1073/pnas.0915133107.

William, J., Euler, C., Christensen, S., & Shlomchik, M. J. (2002). Evolution of autoantibody responses via somatic hypermutation outside of germinal centers. *Science*, *297*(5589), 2066–2070.

Woodruff, M. F., Reid, B., & James, K. (1967). Effect of antilymphocytic antibody and antibody fragments on human lymphocytes in vitro. *Nature*, *215*(5101), 591–594.

Wotherspoon, A. C., Doglioni, C., Diss, T. C., Pan, L., Moschini, A., de Boni, M., et al. (1993). Regression of primary low-grade B-cell gastric lymphoma of mucosa-associated lymphoid tissue type after eradication of Helicobacter pylori. *Lancet*, *342*(8871), 575–577.

Xiong, H., Dolpady, J., Wabl, M., Curotto de Lafaille, M. A., & Lafaille, J. J. (2012). Sequential class switching is required for the generation of high affinity IgE antibodies. *The Journal of Experimental Medicine*, *209*(2), 353–364. http://dx.doi.org/10.1084/jem.20111941.

Yang, J., & Reth, M. (2010a). The dissociation activation model of B cell antigen receptor triggering. *FEBS Letters*, *584*(24), 4872–4877. http://dx.doi.org/10.1016/j.febslet.2010.09.045.

Yang, J., & Reth, M. (2010b). Oligomeric organization of the B-cell antigen receptor on resting cells. *Nature*, *467*(7314), 465–469. http://dx.doi.org/10.1038/nature09357.

Yoshida, K., Matsuoka, M., Usuda, S., Mori, A., Ishizaka, K., & Sakano, H. (1990). Immunoglobulin switch circular DNA in the mouse infected with Nippostrongylus brasiliensis: Evidence for successive class switching from mu to epsilon via gamma 1. *Proceedings of the National Academy of Sciences of the United States of America*, *87*(20), 7829–7833.

Young, R. M., & Staudt, L. M. (2013). Targeting pathological B cell receptor signalling in lymphoid malignancies. *Nature Reviews. Drug Discovery*, *12*(3), 229–243. http://dx.doi.org/10.1038/nrd3937.

CHAPTER TWO

A Critical Role for Cell Polarity in Antigen Extraction, Processing, and Presentation by B Lymphocytes

Dorian Obino, Ana-Maria Lennon-Duménil[1]
INSERM U932, Institut Curie, Paris, France
[1]Corresponding author: e-mail address: ana-maria.lennon@curie.fr

Contents

Abstract

The activation of B lymphocytes in response to external stimuli represents a key step in the adaptive immune response, which is required for the production of high-affinity antibodies and for the generation of long-term memory. Because the dysregulation of B lymphocyte responses can lead to diverse pathological situations, B cells are considered today as valuable therapeutic targets for immunomodulation, in particular in the context of autoimmune reactions. Here, we review the fundamental molecular and cell biological mechanisms that enable B cells to efficiently sense, acquire, and respond to extracellular antigens. A special emphasis is given to cell polarity, which was shown to be critical for the regulation of antigen acquisition, processing, and presentation by B lymphocytes. How cell polarity coordinates the various steps of B lymphocyte activation and might impact the humoral immune response is further discussed.

Advances in Immunology, Volume 123
ISSN 0065-2776
http://dx.doi.org/10.1016/B978-0-12-800266-7.00001-7

1. INTRODUCTION

In the lymph nodes, B cells acquire antigens that are tethered at the surface of neighboring cells such as migrating or follicular dendritic cells or subcapsular sinus (SCS) macrophages (Carrasco & Batista, 2007; Junt et al., 2007; Qi, Egen, Huang, & Germain, 2006; Suzuki, Grigorova, Phan, Kelly, & Cyster, 2009). The recognition by B cells of such immobilized antigens through the engagement of their B-cell antigen receptors (BCRs) results in the establishment of an immune synapse (Batista, Iber, & Neuberger, 2001), which is a key step in the process of B-cell activation (Carrasco, Fleire, Cameron, Dustin, & Batista, 2004; Fleire et al., 2006). Antigen recognition is then followed by a rapid polarization of B cells where both the microtubule-organizing center (MTOC) and lysosomes are repositioned at the immune synapse (Yuseff et al., 2011). Lysosome recruitment and secretion at the synaptic interface promote antigen extraction and processing for further presentation to primed $CD4^+$ T cells (Yuseff et al., 2011). Upon activation, while a pool of B cells differentiates into short-lived plasmablasts producing antibodies with relative low affinity (Cunningham et al., 2007), another pool receives signals from T follicular helper cells. This cell-cell contact — known as T-cell/B-cell cooperation — is required for B cells to be fully activated, proliferate, and form germinal centers (GCs) (Mitchison, 2004). Here, B cells undergo affinity maturation and class switch recombination leading to their differentiation into either plasma cells, which produce high-affinity antibodies, or memory B cells (Allen, Okada, & Cyster, 2007; MacLennan, 1994). Thus, the outcome of the adaptive immune response relies on the ability of B cells to polarize upon antigen stimulation. In this chapter, a special emphasis is given to the crucial role for the spatiotemporal coordination of the membrane-trafficking and polarization events that are required for B cells to efficiently extract, process, and present antigens (Yuseff, Pierobon, Reversat, & Lennon-Duménil, 2013; Yuseff et al., 2011).

2. B-CELL ACTIVATION IN SECONDARY LYMPHOID ORGANS

2.1. Antigen encounter by B cells

Due to the antigenic specificity of each BCR, the probability for a given B cell to encounter its cognate antigen by patrolling the body is very low. In contrast, the follicles of secondary lymphoid organs where B cells reside provide the appropriate environment for them to encounter their specific antigens. Antigens that are present in the peripheral tissues are collected within lymph nodes, which form an "antigen reservoir" for cognate B cells. The fate of antigens in the lymph nodes is mainly dependent on their nature (Batista & Harwood, 2009). Small antigens from afferent lymph vessels diffuse into the lymph node follicle where B cells acquire them in a soluble form (Pape, Catron, Itano, & Jenkins, 2007). The mechanisms by which soluble antigens gain access to B-cell follicles are not totally understood, but it has been proposed that their ingress is promoted by tiny pores (Batista & Harwood, 2009) and/or a conduit network that directly connects the subcapsular sinus (SCS) with B-cell follicles (Roozendaal et al., 2009). In contrast, large antigens (greater than 70 kDa) remain trapped at the SCS floor site where $CD169^+$ macrophages have been shown to capture and transfer them (Carrasco & Batista, 2007; Junt et al., 2007) to follicular B cells. How antigens are transferred from the SCS to B-cell follicles is unclear. Two routes have been suggested: First, the subcapsular $CD169^+$ macrophages that display a poor phagocytic capacity could recycle antigens and expose them in their native form at the cell surface. Second, antigens that are immobilized at the surface of $CD169^+$ macrophages could be translocated from the SCS to B-cell follicles through macrophage protrusions (Martinez-Pomares & Gordon, 2007). In all cases, antigens are presented to cognate B cells and the engagement of the BCR with such surface-tethered antigens triggers the establishment of an immune synapse between the antigen-carrying cell and the B cell.

2.2. The establishment of the immune synapse

The immune synapse formed by B cells after the engagement of their BCR with cell-surface-bound antigens resembles the one originally described in the T cells (Grakoui, 1999; Yuseff, Lankar, & Lennon-Duménil, 2009). The BCR is composed of a cell-surface immunoglobulin associated to a signaling module that comprises the Igα–Igβ dimer. This dimer contains immunoreceptor tyrosine-based activation motifs that are rapidly phosphorylated by Src family kinases upon antigen recognition (Cambier, Pleiman, & Clark, 1994; Reth & Wienands, 1997). This initiates a cascade of signaling events involving cytosolic molecules such as Syk, Lyn, and Vav (Depoil et al., 2009; Tolar, Sohn, & Pierce, 2008) and calcium flux that ultimately launches the transcription of genes required for B-cell function (Baba &

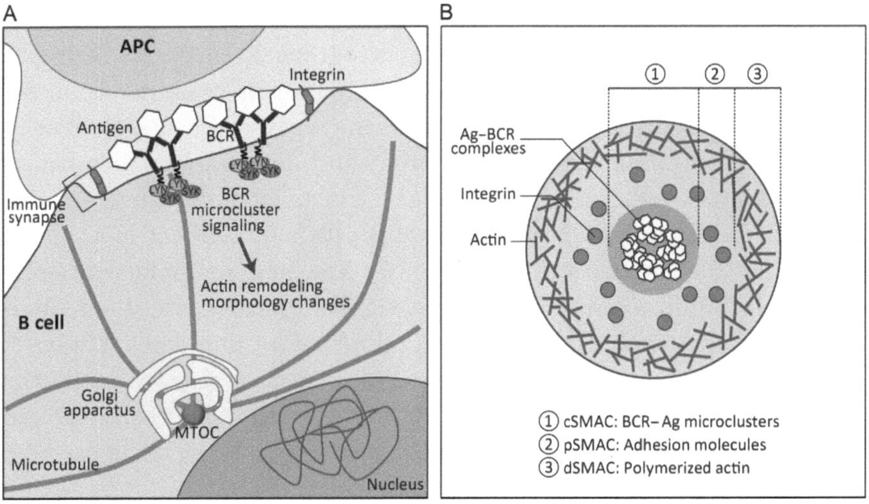

Figure 2.1 B cells form an immune synapse upon BCR engagement with immobilized antigens. (A) The engagement of the BCR with cell surface-tethered antigens leads to signaling events that trigger profound actin cytoskeleton remodeling: The B cell spreads on the antigen-presenting cell and the reorganization of its membrane at the antigen-contact site forms an immune synapse. (B) The immune synapse is a highly organized platform that includes three concentric areas: At the synapse center, antigen–BCR microclusters concentrate and form the central supramolecular activation cluster or cSMAC ①. This is surrounded by the peripheral SMAC where adhesion molecules such as LFA-1 localize ② and the distal SMAC where polymerized actin accumulates ③. (See the color plate.)

Kurosaki, 2011) (Fig. 2.1A). Concomitantly, drastic modifications of the B-cell actin cytoskeleton take place. First, BCR stimulation initiates a rapid cofilin-dependent actin cytoskeleton depolymerization at the antigen-contact site, allowing the local increase of BCR diffusion within the membrane (Freeman et al., 2011). This is followed by polarized actin repolymerization that promotes B-cell spreading onto the antigen-carrying cell. This process facilitates the formation of BCR microclusters that are required for sustained BCR signaling (Treanor et al., 2010). Then, BCR microclusters merge together in the center of the synapse to form the central supramolecular activation cluster (cSMAC) through a contraction phase mediated by myosin II (Fleire et al., 2006), which requires the coordinated action of the ezrin, radixin, and moesin proteins (Treanor, Depoil, Bruckbauer, & Batista, 2011) and the microtubule minus-end molecular motor dynein (Schnyder et al., 2011). Cortical actin rearrangements also trigger the formation of a second concentric region called the peripheral SMAC (pSMAC) that contains adhesion molecules such as LFA-1 (lymphocyte function-associated antigen 1, also known as $\alpha_L\beta_2$ integrin). The engagement of this integrin with its counterreceptor intercellular adhesion molecule 1 (also known as CD54) promotes B-cell adhesion to the antigen-presenting cell, facilitating the formation of the immune synapse and decreasing the antigen-affinity threshold required for BCR-mediated B-cell activation (Carrasco et al., 2004). Whereas the cSMAC and the pSMAC display very low concentration of polymerized actin, actin concentrates in a third concentric region called the distal SMAC (dSMAC) (Fig. 2.1B). Thus, reorganization of the actin cytoskeleton upon BCR engagement with surface-tethered antigen is a crucial stage in the early events of B-cell responses. In addition, the establishment of the immune synapse gives the first cues of asymmetrical cell organization leading *in fine* to the polarization of B-cell organelles towards the synaptic interface.

3. B-CELL POLARITY AND ANTIGEN PROCESSING AND PRESENTATION

3.1. B-cell polarization upon particulate antigen stimulation

Similarly to observations that were made in natural killer and cytotoxic T cells (Stinchcombe et al., 2011), upon BCR engagement, B cells rapidly polarize their MTOC towards the antigen-contact site. Concomitantly, MHC class II^+/Lamp-1^+ lysosomes are recruited to the immune synapse, where they cluster (Yuseff et al., 2011) (Fig. 2.2A). Interestingly, using MHC class II-expressing human melanoma cells, Wubbolts et al. had found that MHC class II^+ lysosomes traffic along microtubules to reach the plasma membrane (Wubbolts et al., 1996, 1999). Accordingly, we have shown that laser ablation of the MTOC following B-cell stimulation prevents polarized recruitment of lysosomes at the immune synapse (Yuseff et al., 2011). Although the molecular mechanisms underlying B-cell polarization upon BCR engagement with surface-tethered antigens are not fully understood, we identified the conserved polarity machinery cell division control protein 42 (Cdc42) and the atypical protein kinase C zeta-type (aPKC-ζ) (Yuseff et al., 2013) as required for B-cell polarization. Silencing of either Cdc42 or aPKC-ζ impairs MTOC and lysosome relocation at the synaptic interface, giving rise to a drastic decrease in antigen extraction, processing, and presentation (Yuseff et al., 2011).

B-cell polarization is not restricted to the MTOC and lysosomes. Indeed, we have found that upon BCR engagement with surface-tethered antigens, B cells polarize their Golgi apparatus and reorient their nucleus so that a

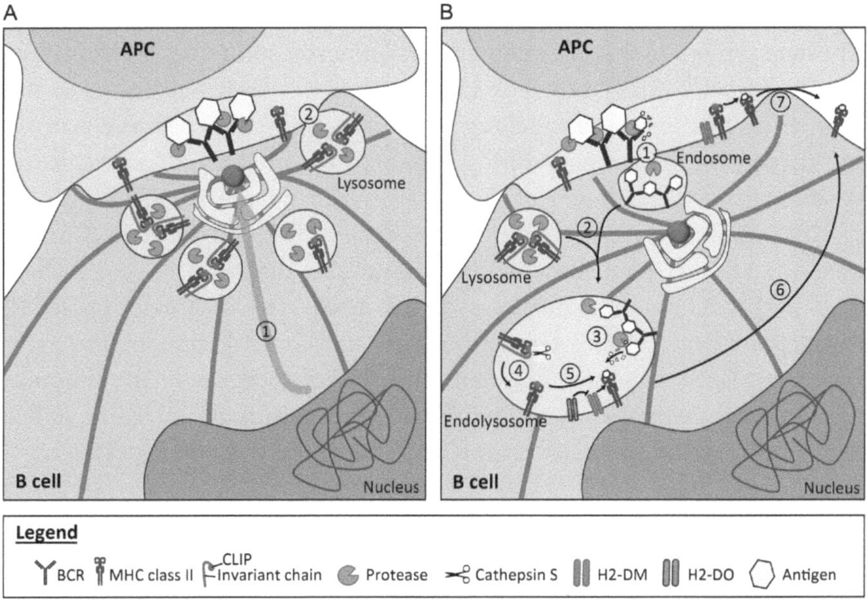

Figure 2.2 B cells polarize upon antigen stimulation. (A) Antigen stimulation initiates morphological changes within B cells. The microtubule-organizing center (MTOC) together with the Golgi apparatus reposition at the immune synapse ①. Concomitantly, lysosomes that were dispersed at the cell periphery cluster around the MTOC and are recruited at the antigen-contact site where they are secreted ②, resulting in synapse acidification (light pink) and the local release of proteases. (B) Synapse acidification and protease release promote efficient antigen extraction within the synaptic space leading to antigen internalization in B cell–endosomes ①. Endolysosomes that are formed by the fusion ② of antigen-containing endosomes with lysosomes that carry MHC class II molecules allow the efficient processing of antigens ③. In the same compartment, cathepsin S cleaves the invariant chain (Ii) ④ resulting in MHC class II–CLIP complex formation. Finally, H2-DM molecules promote the exchange between CLIP and antigenic peptides for them to be loaded onto MHC class II molecules ⑤. The catalysis of CLIP released by H2-DM is regulated by another nonclassical MHC II molecule, H2-DO. Peptide–MHC class II complexes are exported to the B-cell surface ⑥. It is important to note that antigenic peptides might also be generated within the synaptic space where they are directly loaded onto MHC class II molecules at the cell surface ⑦. (See the color plate.)

nucleus invagination faces the immune synapse (unplublished data and Fig. 2.2A). Interestingly, it has been recently described that during B-cell development, nuclear positioning of genes undergoes global changes that orchestrates B-cell fate (Lin et al., 2012). Taken together, it is tempting to speculate that antigenic stimulation leads to a global B-cell reorganization through the acquisition of polarity. In this context, BCR engagement with surface-tethered antigens might coordinate (1) the polarization of the MTOC at the immune synapse to guide lysosome recruitment at the synaptic interface, (2) the repositioning of the Golgi apparatus in close proximity of the synapse to facilitate the export of neosynthetized proteins, and (3) the remodeling of chromatin to allow the transcription of genes required for B-cell function. How these events are coordinated in time and space remains to be determined. Is B-cell polarization established in a sequential manner? Does BCR stimulation initiate a "program" of global morphological changes that leads to the independent polarization of organelles? As discussed in the preceding text, we have shown that lysosome clustering at the immune synapse relies on the repositioning of the MTOC (Yuseff et al., 2011), providing the first clue on the interdependence of the polarization events observed upon BCR engagement.

3.2. Antigen extraction from antigen-presenting cell membranes

The immune synapse that forms at the interface between antigen-carrying cells and B lymphocytes corresponds to a dynamic structure where both endocytic and exocytic events take place. In T cells, it has been shown that effector molecules such as cytokines and lytic granules are secreted through the immune synapse towards target cells (Poo, Conrad, & Janeway, 1988; Stinchcombe et al., 2011). Exocytosis at the T-cell synapse occurs in specific domains that display very low levels of polymerized actin (Stinchcombe, Majorovits, Bossi, Fuller, & Griffiths, 2006), facilitating the local secretion of molecules (Chemin et al., 2012).

In B cells, upon BCR engagement with immobilized antigens, lysosomes are recruited at the immune synapse where they are locally secreted, thereby leading to the acidification of the extracellular synaptic space and the release of proteases in this confined environment (Yuseff et al., 2011) (Fig. 2.2A). Both synapse alkalinization and extracellular inhibition of proteases lead to a drastic decrease in the capacity of B cells to extract antigens, indicating that synapse acidification and protease secretion are required for efficient extraction of immobilized antigens (Yuseff et al., 2011). Whether exocytic events at the B-cell synapse specifically take place in areas displaying low concentration of polymerized actin has not been addressed so far.

The integration of mechanics into this biological process has recently been proposed by showing that B cells use myosin IIA-mediated pulling forces to discriminate between membrane-bound antigens displaying different affinities for the BCR (Natkanski et al., 2013). Using flexible and fluid plasma membrane sheets that display mechanical properties close to those of dendritic cells, they have shown that upon BCR engagement, myosin II pulls on the presenting membrane leading to its invagination and subsequent antigen internalization into clathrin-coated pits. The lifetime of membrane invaginations and the probability of antigen internalization depend on the strength of antigen–BCR interactions. Using antigens with different affinities for the BCR, they further showed that myosin IIA activity impairs the internalization of relatively low-affinity antigens by disrupting antigen–BCR bonds. Myosin IIA activity is therefore critical for B cells to discriminate antigen affinities and later develop productive antibody responses (Natkanski et al., 2013). This work further raises the question of whether myosin II-dependent mechanical forces are generally required for endocytosis of immobilized molecules, what might be frequent *in vivo*. Interestingly, it has been reported that during epithelial morphogenesis in the *Drosophila* embryo, myosin IIA is asymmetrically distributed within cells where it concentrates at anterior–posterior boundaries of cells. Polarized actomyosin contractility in the ventrolateral region of the epithelium destabilizes the adherens junctions by promoting local clathrin-dependent endocytosis of E-cadherin and defines the axis of cell division (Levayer, Pelissier-Monier, & Lecuit, 2011). However, the mechanism(s) underlying myosin IIA asymmetrical distribution within cells and its anchorage at the plasma membrane remain elusive.

Taken together, these studies suggest that both proteases that are exocytosed into the synaptic space and myosin IIA-mediated pulling forces help in internalizing antigens at the B-cell synapse. Whether mechanical forces exerted by myosin IIA contractility are sufficient to break pieces of the presenting membrane leading to their internalization together with antigens remains to be precisely determined. In addition, the role of lysosomal lipases that are secreted at the immune synapse and might help in acquiring membrane-bound antigens should be investigated. We propose that the requirement for extracellular proteolysis and/or mechanical forces to internalize antigens into B cells might depend on their biochemical and physical properties.

3.3. Processing of antigen–BCR complexes

Antigenic peptides are generated from extracted antigens and loaded onto MHC class II molecules for presentation to primed $CD4^+$ T cells (Fig. 2.2B). This step — known as T-cell/B-cell cooperation — is pivotal for the ultimate formation of GCs and production of high-affinity antibodies by B lymphocytes (Mitchison, 2004).

BCR-dependent activation of the cytoplasmic tyrosine kinase Syk is required for actin cytoskeleton reorganization and efficient antigen–BCR trafficking to MHC class II$^+$ endolysosomes (Le Roux et al., 2007) (Fig. 2.2B ① and ②). There, antigens undergo limited proteolysis in order to preserve T-cell epitopes from excessive degradation (Delamarre, Pack, Chang, Mellman, & Trombetta, 2005) (Fig. 2.2B ③). Interestingly, in addition to its role in the uptake of immobilized antigens, myosin IIA was also shown to facilitate the arrival of antigen–BCR complexes into MHC class II$^+$ endolysosomes through its interaction with the cytosolic tail of invariant chain (Ii) (Vascotto et al., 2007). Ii associates to MHC class II molecules during biogenesis in the endoplasmic reticulum, preventing the premature binding of endogenous peptides onto MHC class II molecules (Bakke & Dobberstein, 1990; Lotteau et al., 1990; Roche & Cresswell, 1990, 1991). Once in endolysosome, Ii undergoes sequential proteolysis by aspartic and cysteine proteases including cathepsin S and the intramembrane endoprotease SPPL2a. This ultimately leads to the generation of the Ii CLIP fragment that occupies the MHC class II peptide-binding groove (Driessen et al., 1999; Riese et al., 1996; Villadangos, Riese, Peters, Chapman, & Ploegh, 1997) (Fig. 2.2B ④). Noticeably, SPPL2a knockout mice display a defect in B-cell development due to intracellular accumulation of Ii, suggesting that the last step of Ii cleavage is a checkpoint during B-cell differentiation (Beisner et al., 2013; Schneppenheim et al., 2013). CLIP exchange with antigenic peptides is catalyzed by the nonclassical MHC class II molecule, H2-DM (Denzin & Cresswell, 1995) (Fig. 2.2B ⑤). B cells express an additional nonclassical MHC class II molecule, H2-DO, which was shown to downregulate the catalysis of CLIP released by H2-DM (Denzin, 1997). Accordingly, H2-DO knockout B cells were found to exhibit increased amount of MHC class II–peptide complexes and compete wild-type B cells for the entry to GCs (Draghi & Denzin, 2010). Finally, peptide–MHC class II complexes are exported to the B-cell surface for further presentation to primed CD4$^+$ T cells (Fig. 2.2B ⑥). Remarkably, antigen processing and peptide loading onto MHC class II molecules can also directly take place at the B-cell surface, where H2-DM molecules are equally found (Moss, Tree, & Watts, 2007) (Fig. 2.2B ⑦).

3.4. Antigen presentation to helper T cells

Following antigen internalization and processing, B cells migrate towards the T-cell boundary, where they establish contacts with cognate T cells (Pereira, Kelly, & Cyster, 2010). B-cell migration is supported by the upregulation of CCR7 (CC chemokine receptor 7) that senses CCL19 and CCL21

(CC chemokine ligands 19 and 21) produced by stromal cells in the T-cell zone (Reif et al., 2002). While migrating towards the T-cell zone, B cells concentrate internalized antigens in their uropod (Carrasco & Batista, 2007), suggesting that they maintain the polarized phenotype acquired upon BCR engagement with immobilized antigens. This polarized concentration of antigens leads to their asymmetrical distribution among daughter B cells following cell division, providing them with differential antigen presentation capacities (Thaunat et al., 2012). Noticeably, the contact between B cells and T cells also results in the establishment of an immune synapse, where both cells harbor polarized phenotypes (Duchez, Rodrigues, Bertrand, & Valitutti, 2011). At the synapse, bidirectional exchanges take place, ultimately leading to the formation of GCs where B cells undergo affinity maturation and class switch recombination and differentiate into either plasma cells producing high-affinity antibodies or long-lived memory B cells (Allen et al., 2007; MacLennan, 1994). Interestingly, asymmetrical polarized division of B cells was also observed during the GC reaction and shown to regulate the survival of daughter cells through the unequal inheritance of Bcl-6, IL-21 receptor, and the polarity protein aPKC-ζ (Barnett et al., 2011). Thus, cell polarity is likely to play an essential role in shaping B-cell responses to surface-tethered antigens at their various activation stages. Whether the initial B-cell polarity established upon BCR engagement with immobilized antigens impacts on asymmetrical cell division and formation of the immune synapse with T cells shall now be investigated.

4. EXTRACELLULAR CUES THAT TUNE B-CELL POLARITY AND ACTIVATION

4.1. Chemokines involved in follicle patrolling by B cells

It is clear that the microenvironment of lymphoid tissues plays a key role in (1) modulating the ability of B cells to respond to antigens, (2) driving B-cell differentiation and fate, and (3) tuning the overall outcome of the adaptive immune response. Whereas numerous studies have focused their interest in understanding how the lymphoid microenvironment impacts B-cell homing within lymph nodes or the GC reaction, only few data are available on how it modifies the ability of B cells to initially respond to surface-bound antigens. Efficient antigen sampling by B cells relies on their capability to explore entire follicles within lymph nodes. This is achieved in response to the chemokine CXCL13 that induces CXCR5-expressing B cells to continuously migrate by random walking (Miller, Wei, Parker, & Cahalan, 2002; Sáez de Guinoa, Barrio, Mellado, & Carrasco, 2011). B-cell spreading onto antigen-presenting cells and immune synapse establishment are coupled to an arrest phase in B-cell migration during which B cells acquire antigens (Carrasco & Batista, 2007; Junt et al., 2007). Thereby, the tight regulation of the interplay between cell motility and antigen internalization is critical for B-cell function. It has been recently shown that recruitment to the immune synapse of the scaffold protein vinculin, which links integrins at the plasma membrane with the actin cytoskeleton (Humphries et al., 2007), regulates adhesion between the B cell and the antigen-presenting cell (Saez de Guinoa, Barrio, & Carrasco, 2013). Vinculin is recruited to the immune synapse, in parallel to a phosphatidylinositol (4,5)-bisphosphate (PIP_2) wave, and stabilizes LFA-1 cluster within the pSMAC. This process relies on the activity of the myosin IIA motor, which is also required for CXCL13-mediated migration (Sáez de Guinoa et al., 2011). Vinculin and myosin IIA are therefore critical for B lymphocytes to switch between random motility and antigen internalization. Thus, B cells possess the machinery required to (1) integrate incoming information and (2) adapt their response to maximize antigen acquisition by regulating the balance between cell migration and cell adhesion to antigen-presenting cells. This leads to the formation either of a stable immune synapse characterized by a firm adhesion of the B cell to the antigen-presenting cell and very poor migratory capacities or of CXCL13- and LFA-1-mediated kinapses where B cells establish successive short interactions with multiple antigen-presenting cells, alternating between adhesion and motility phases. Interestingly, this resembles the intermittent search strategy displayed by dendritic cells where the coupling between antigen processing and cell motility by myosin II was proposed to enable dendritic cells to efficiently patrol peripheral tissues in search for antigens (Faure-andré et al., 2008; Heuzé et al., 2013). Myosin II activity therefore emerges as pivotal for many steps of B-cell activation: (1) It stabilizes vinculin by promoting LFA-1-mediated B-cell spreading onto the antigen-presenting cell and might thereby decrease B-cell velocity, (2) it provides a mechanical force to "test" for discrimination of antigen affinity, and (3) it ensures the proper trafficking of antigen–BCR complexes into MHC class II^+ lysosomes for antigen processing. Whether myosin II regulates B-cell polarity and whether B-cell polarity contributes to the switch between antigen uptake and cell migration remain to be addressed.

4.2. Galectins and B-cell function

Beyond chemokines, B-cell follicles contain a diversity of stromal cells and extracellular matrix components that might modulate the outcome of

B lymphocyte responses. Such cues have been poorly studied. During the last decade, glycan-binding proteins have emerged as key regulators of immune cell homeostasis and response to antigens (Rabinovich & Croci, 2012). Among these proteins is the galectin family (Rabinovich & Toscano, 2009) that has the ability to cross-link cell-surface glycol proteins in the extracellular space upon secretion, thereby impacting a wide range of biological processes. For instance, it was described that the lack of galectin-3 in *Schistosoma mansoni* chronically infected mice promotes plasma cell formation (Oliveira et al., 2011), suggesting a role for this galectin in negative regulation of B-cell responses. In contrast, galectin-1 and galectin-8 were shown to bind to mature B cells and promote the differentiation of LPS-treated B cells into antibody-secreting plasma cells *in vitro* (Tsai et al., 2008, 2011). It is interesting to note that while galectin-1 expression increases during the course of B-cell differentiation upon LPS stimulation, the expression of galectin-8 decreases (Tsai et al., 2011), suggesting an important role for galectin-8 in the early steps of B-cell activation. In addition, high concentration of galectin-8 was reported to trigger antigen-independent proliferation of $CD4^+$ T cells, whereas lower quantity of the protein provides costimulatory signals that synergize antigen-specific $CD4^+$ T-cell responses (Cattaneo, Tribulatti, & Campetella, 2011; Tribulatti, Cattaneo, Hellman, Mucci, & Campetella, 2009). Interestingly, galectin-8 was also described to promote cell adhesion (Cueni & Detmar, 2009) and migration of endothelial cells *in vivo* (Delgado et al., 2011). Moreover, the presence of function-blocking autoantibodies against galectin-8 in the sera of patients suffering from systemic lupus erythematosus correlates with acute lymphopenia (Massardo et al., 2009), suggesting that galectin-8 binding to glycosylated proteins might regulate B-cell homeostasis and function. Whether galectins or other glycan-binding proteins that are present within the lymphoid environment modulate specific steps of B-cell activation such as BCR signaling, B-cell adhesion, polarity, or migration remains to be explored.

5. CONCLUDING REMARKS

Although the signaling and genetic processes that enable B lymphocytes to form GCs and differentiate into long-lived plasma cells have been extensively studied in the past, the fundamental cell biological events that drive B-cell activation in response to antigen stimulation have recently started to be unraveled. Here, we have reviewed the fundamental molecular and cell biological mechanisms involved in B-cell function and further highlighted how the acquisition of cell polarity critically regulates the various stages of B-cell responses to surface-tethered antigens. How B-cell polarity is controlled by extracellular cues in the context of lymphoid organs and whether the modulation of B-cell polarity can be used to tune humoral immune responses in pathological situations shall now be investigated.

ACKNOWLEDGMENTS

The authors acknowledge Maria-Isabel Yuseff, Anne Reversat, Paolo Pierobon, Odile Malbec, and Danielle Lankar for critical discussions. D. O. was supported by a PhD fellowship from the Ecole Doctorale B3MI, Université Paris Diderot. Funding was obtained from the Association Nationale pour la Recherche (ANR-PoLyBex-12-BSV3-0014-001) and the European Research Council (ERC-Strapacemi-GA 243103) to A- M. L- D.

REFERENCES

Allen, C. D. C., Okada, T., & Cyster, J. G. (2007). Germinal-center organization and cellular dynamics. *Immunity*, *27*(2), 190–202. http://dx.doi.org/10.1016/j.immuni.2007.07.009.

Baba, Y., & Kurosaki, T. (2011). Impact of Ca2+ signaling on B cell function. *Trends in Immunology*, *32*(12), 589–594. http://dx.doi.org/10.1016/j.it.2011.09.004.

Bakke, O., & Dobberstein, B. (1990). MHC class II-associated invariant chain contains a sorting signal for endosomal compartments. *Cell*, *63*(4), 707–716. Retrieved from, http://www.ncbi.nlm.nih.gov/pubmed/2121367.

Barnett, B. E., Ciocca, M. L., Goenka, R., Barnett, L. G., Wu, J., Laufer, T. M., et al. (2011). Asymmetric B cell division in the germinal center reaction. *Science (New York, N.Y.)*, *335*, 342–344. http://dx.doi.org/10.1126/science.1213495.

Batista, F. D., & Harwood, N. E. (2009). The who, how and where of antigen presentation to B cells. *Nature Reviews. Immunology*, *9*(1), 15–27. http://dx.doi.org/10.1038/nri2454.

Batista, F. D., Iber, D., & Neuberger, M. S. (2001). B cells acquire antigen from target cells after synapse formation. *Nature*, *411*(6836), 489–494. http://dx.doi.org/10.1038/35078099.

Beisner, D. R., Langerak, P., Parker, A. E., Dahlberg, C., Otero, F. J., Sutton, S. E., et al. (2013). The intramembrane protease Sppl2a is required for B cell and DC development and survival via cleavage of the invariant chain. *The Journal of Experimental Medicine*, *210*(1), 23–30. http://dx.doi.org/10.1084/jem.20121072.

Cambier, J. C., Pleiman, C. M., & Clark, M. R. (1994). Signal transduction by the B cell antigen receptor and its coreceptors. *Annual Review of Immunology*, *12*, 457–486.

Carrasco, Y. R., & Batista, F. D. (2007). B cells acquire particulate antigen in a macrophage-rich area at the boundary between the follicle and the subcapsular sinus of the lymph node. *Immunity*, *27*(1), 160–171. http://dx.doi.org/10.1016/j.immuni.2007.06.007.

Carrasco, Y. R., Fleire, S. J., Cameron, T., Dustin, M. L., & Batista, F. D. (2004). LFA-1/ICAM-1 interaction lowers the threshold of B cell activation by facilitating B cell adhesion and synapse formation. *Immunity*, *20*(5), 589–599. Retrieved from, http://www.ncbi.nlm.nih.gov/pubmed/15142527.

Cattaneo, V., Tribulatti, M. V., & Campetella, O. (2011). Galectin-8 tandem-repeat structure is essential for T-cell proliferation but not for co-stimulation. *The Biochemical Journal*, *434*(1), 153–160. http://dx.doi.org/10.1042/BJ20101691.

Chemin, K., Bohineust, A., Dogniaux, S., Tourret, M., Guégan, S., Miro, F., et al. (2012). Cytokine secretion by CD4+ T cells at the immunological synapse requires Cdc42-dependent local actin remodeling but not microtubule organizing center polarity. *Journal of Immunology (Baltimore, Md: 1950)*, *189*(5), 2159–2168. http://dx.doi.org/10.4049/jimmunol.1200156.

Cueni, L. N., & Detmar, M. (2009). Galectin-8 interacts with podoplanin and modulates lymphatic endothelial cell functions. *Experimental Cell Research*, *315*(10), 1715–1723. http://dx.doi.org/10.1016/j.yexcr.2009.02.021.

Cunningham, A. F., Gaspal, F., Serre, K., Mohr, E., Henderson, I. R., Scott-Tucker, A., et al. (2007). Salmonella induces a switched antibody response without germinal centers that impedes the extracellular spread of infection. *Journal of Immunology (Baltimore, Md: 1950)*, *178*(10), 6200–6207. Retrieved from, http://www.ncbi.nlm.nih.gov/pubmed/17475847.

Delamarre, L., Pack, M., Chang, H., Mellman, I., & Trombetta, E. S. (2005). Differential lysosomal proteolysis in antigen-presenting cells determines antigen fate. *Science (New York, N.Y.)*, *307*(5715), 1630–1634. http://dx.doi.org/10.1126/science.1108003.

Delgado, V. M. C., Nugnes, L. G., Colombo, L. L., Troncoso, M. F., Fernández, M. M., Malchiodi, E. L., et al. (2011). Modulation of endothelial cell migration and angiogenesis: A novel function for the "tandem-repeat" lectin galectin-8. *FASEB Journal: Official Publication of the Federation of American Societies for Experimental Biology*, *25*(1), 242–254. http://dx.doi.org/10.1096/fj.09-144907.

Denzin, L. K. (1997). Negative regulation by HLA-DO of MHC class II-restricted antigen processing. *Science*, *278*(5335), 106–109. http://dx.doi.org/10.1126/science.278.5335.106.

Denzin, L. K., & Cresswell, P. (1995). HLA-DM induces CLIP dissociation from MHC class II alpha beta dimers and facilitates peptide loading. *Cell*, *82*(1), 155–165. Retrieved from, http://www.ncbi.nlm.nih.gov/pubmed/7606781.

Depoil, D., Weber, M., Treanor, B., Fleire, S. J., Carrasco, Y. R., Harwood, N. E., et al. (2009). Early events of B cell activation by antigen. *Science Signaling*, *2*(63), pt1. http://dx.doi.org/10.1126/scisignal.263pt1.

Draghi, N. A., & Denzin, L. K. (2010). H2-O, a MHC class II-like protein, sets a threshold for B-cell entry into germinal centers. *Proceedings of the National Academy of Sciences of the United States of America*, *107*(38), 16607–16612. http://dx.doi.org/10.1073/pnas.1004664107.

Driessen, C., Bryant, R. A., Lennon-Duménil, A. M., Villadangos, J. A., Bryant, P. W., Shi, G. P., et al. (1999). Cathepsin S controls the trafficking and maturation of MHC class II molecules in dendritic cells. *The Journal of Cell Biology*, *147*(4), 775–790. Retrieved from, http://www.pubmedcentral.nih.gov/articlerender.fcgi?artid=2156161&tool=pmcentrez&rendertype=abstract.

Duchez, S., Rodrigues, M., Bertrand, F., & Valitutti, S. (2011). Reciprocal polarization of T and B cells at the immunological synapse. *Journal of Immunology (Baltimore, Md: 1950)*, *187*(9), 4571–4580. http://dx.doi.org/10.4049/jimmunol.1100600.

Faure-andré, G., Vargas, P., Yuseff, M., Heuzé, M., Vascotto, F., Boulanger, J., et al. (2008). Regulation of dendritic cell migration by CD74, the MHC class II-associated Invariant chain. *Science*, *322*, 1705–1710.

Fleire, S. J., Goldman, J. P., Carrasco, Y. R., Weber, M., Bray, D., & Batista, F. D. (2006). B cell ligand discrimination through a spreading and contraction response. *Science (New York, N.Y.)*, *312*(5774), 738–741. http://dx.doi.org/10.1126/science.1123940.

Freeman, S. A., Lei, V., Dang-Lawson, M., Mizuno, K., Roskelley, C. D., & Gold, M. R. (2011). Cofilin-mediated F-actin severing is regulated by the Rap GTPase and controls the cytoskeletal dynamics that drive lymphocyte spreading and BCR microcluster formation. *Journal of Immunology (Baltimore, Md: 1950)*, *187*(11), 5887–5900. http://dx.doi.org/10.4049/jimmunol.1102233.

Grakoui, A. (1999). The immunological synapse: A molecular machine controlling T cell activation. *Science*, *285*(5425), 221–227. http://dx.doi.org/10.1126/science.285.5425.221.

Heuzé, M. L., Vargas, P., Chabaud, M., Le Berre, M., Liu, Y.-J., Collin, O., et al. (2013). Migration of dendritic cells: Physical principles, molecular mechanisms, and functional implications. *Immunological Reviews*, *256*(1), 240–254. http://dx.doi.org/10.1111/imr.12108.

Humphries, J. D., Wang, P., Streuli, C., Geiger, B., Humphries, M. J., & Ballestrem, C. (2007). Vinculin controls focal adhesion formation by direct interactions with talin and actin. *The Journal of Cell Biology*, *179*(5), 1043–1057. http://dx.doi.org/10.1083/jcb.200703036.

Junt, T., Moseman, E. A., Iannacone, M., Massberg, S., Lang, P. A., Boes, M., et al. (2007). Subcapsular sinus macrophages in lymph nodes clear lymph-borne viruses and present them to antiviral B cells. *Nature*, *450*(7166), 110–114. http://dx.doi.org/10.1038/nature06287.

Le Roux, D., Lankar, D., Yuseff, M., Vascotto, F., Yokozeki, T., Faure-andre, G., et al. (2007). Syk-dependent actin dynamics regulate endocytic trafficking and processing of antigens internalized through the B-cell receptor. *Molecular Biology of the Cell*, *18*(September), 3451–3462. http://dx.doi.org/10.1091/mbc.E06.

Levayer, R., Pelissier-Monier, A., & Lecuit, T. (2011). Spatial regulation of Dia and Myosin-II by RhoGEF2 controls initiation of E-cadherin endocytosis during epithelial morphogenesis. *Nature Cell Biology*, *13*(5), 529–540. http://dx.doi.org/10.1038/ncb2224.

Lin, Y. C., Benner, C., Mansson, R., Heinz, S., Miyazaki, K., Miyazaki, M., et al. (2012). Global changes in the nuclear positioning of genes and intra- and interdomain genomic interactions that orchestrate B cell fate. *Nature Immunology*, *13*(12), 1196–1204. http://dx.doi.org/10.1038/ni.2432.

Lotteau, V., Teyton, L., Peleraux, A., Nilsson, T., Karlsson, L., Schmid, S. L., et al. (1990). Intracellular transport of class II MHC molecules directed by invariant chain. *Nature*, *348*, 600–605.

MacLennan, I. C. (1994). Germinal centers. *Annual Review of Immunology*, *12*, 117–139. http://dx.doi.org/10.1146/annurev.iy.12.040194.001001.

Martinez-Pomares, L., & Gordon, S. (2007). Antigen presentation the macrophage way. *Cell*, *131*(4), 641–643. http://dx.doi.org/10.1016/j.cell.2007.10.046.

Massardo, L., Metz, C., Pardo, E., Mezzano, V., Babul, M., Jarpa, E., et al. (2009). Autoantibodies against galectin-8: Their specificity, association with lymphopenia in systemic lupus erythematosus and detection in rheumatoid arthritis and acute inflammation. *Lupus*, *18*(6), 539–546. http://dx.doi.org/10.1177/0961203308099973.

Miller, M. J., Wei, S. H., Parker, I., & Cahalan, M. D. (2002). Two-photon imaging of lymphocyte motility and antigen response in intact lymph node. *Science (New York, N.Y.)*, *296*(5574), 1869–1873. http://dx.doi.org/10.1126/science.1070051.

Mitchison, N. A. (2004). T-cell-B-cell cooperation. *Nature Reviews. Immunology*, *4*(4), 308–312. http://dx.doi.org/10.1038/nri1334.

Moss, C. X., Tree, T. I., & Watts, C. (2007). Reconstruction of a pathway of antigen processing and class II MHC peptide capture. *The EMBO Journal*, *26*(8), 2137–2147. http://dx.doi.org/10.1038/sj.emboj.7601660.

Natkanski, E., Lee, W. Y., Mistry, B., Casal, A., Molloy, J. E., & Tolar, P. (2013). B cells use mechanical energy to discriminate antigen affinities. *Science (New York, N.Y.)*, *340*(6140), 1587–1590. http://dx.doi.org/10.1126/science.1237572.

Oliveira, F. L., Brand, C., Paula, A. A., Arcanjo, K. D., Hsu, D. K., Liu, F.-T., et al. (2011). Lack of galectin-3 disturbs mesenteric lymph node homeostasis and B cell niches in the course of Schistosoma mansoni infection. *PloS One*, *6*(5), e19216. http://dx.doi.org/10.1371/journal.pone.0019216.

Pape, K. A., Catron, D. M., Itano, A. A., & Jenkins, M. K. (2007). The humoral immune response is initiated in lymph nodes by B cells that acquire soluble antigen directly in the follicles. *Immunity*, *26*(4), 491–502. http://dx.doi.org/10.1016/j.immuni.2007.02.011.

Pereira, J. P., Kelly, L. M., & Cyster, J. G. (2010). Finding the right niche: B-cell migration in the early phases of T-dependent antibody responses. *International Immunology*, *22*(6), 413–419. http://dx.doi.org/10.1093/intimm/dxq047.

Poo, W. J., Conrad, L., & Janeway, C. A. (1988). Receptor-directed focusing of lymphokine release by helper T cells. *Nature*, *332*, 378–380.

Qi, H., Egen, J. G., Huang, A. Y. C., & Germain, R. N. (2006). Extrafollicular activation of lymph node B cells by antigen-bearing dendritic cells. *Science (New York, N.Y.)*, *312*(5780), 1672–1676. http://dx.doi.org/10.1126/science.1125703.

Rabinovich, G. A., & Croci, D. O. (2012). Regulatory circuits mediated by lectin-glycan interactions in autoimmunity and cancer. *Immunity*, *36*(3), 322–335. http://dx.doi.org/10.1016/j.immuni.2012.03.004.

Rabinovich, G. A., & Toscano, M. A. (2009). Turning "sweet" on immunity: Galectin-glycan interactions in immune tolerance and inflammation. *Nature Reviews. Immunology*, *9*(5), 338–352. http://dx.doi.org/10.1038/nri2536.

Reif, K., Ekland, E. H., Ohl, L., Nakano, H., Lipp, M., Förster, R., et al. (2002). Balanced responsiveness to chemoattractants from adjacent zones determines B-cell position. *Nature*, *416*(6876), 94–99. http://dx.doi.org/10.1038/416094a.

Reth, M., & Wienands, J. (1997). Initiation and processing of signals from the B cell antigen receptor. *Annual Review of Immunology*, *15*, 453–479.

Riese, R. J., Wolf, P. R., Brömme, D., Natkin, L. R., Villadangos, J. A., Ploegh, H. L., et al. (1996). Essential role for cathepsin S in MHC class II-associated invariant chain processing and peptide loading. *Immunity*, *4*(4), 357–366. Retrieved from, http://www.ncbi.nlm.nih.gov/pubmed/8612130.

Roche, P. A., & Cresswell, P. (1990). Invariant chain association with HLA-DR molecules inhibits immunogenic peptide binding. *Nature*, *345*, 615–618.

Roche, P. A., & Cresswell, P. (1991). Proteolysis of the class II-associated invariant chain generates a peptide binding site in intracellular HLA-DR molecules. *Proceedings of the National Academy of Sciences of the United States of America*, *88*(8), 3150–3154. Retrieved from, http://www.ncbi.nlm.nih.gov/pubmed/21772034.

Roozendaal, R., Mempel, T. R., Pitcher, L. A., Gonzalez, S. F., Verschoor, A., Mebius, R. E., et al. (2009). Conduits mediate transport of low-molecular-weight antigen to lymph node follicles. *Immunity*, *30*(2), 264–276. http://dx.doi.org/10.1016/j.immuni.2008.12.014.

Saez de Guinoa, J., Barrio, L., & Carrasco, Y. R. (2013). Vinculin arrests motile B cells by stabilizing integrin clustering at the immune synapse. *Journal of Immunology (Baltimore, Md: 1950)*, *191*(5), 2742–2751. http://dx.doi.org/10.4049/jimmunol.1300684.

Sáez de Guinoa, J., Barrio, L., Mellado, M., & Carrasco, Y. R. (2011). CXCL13/CXCR5 signaling enhances BCR-triggered B-cell activation by shaping cell dynamics. *Blood*, *118*(6), 1560–1569. http://dx.doi.org/10.1182/blood-2011-01-332106.

Schneppenheim, J., Dressel, R., Hüttl, S., Lüllmann-Rauch, R., Engelke, M., Dittmann, K., et al. (2013). The intramembrane protease SPPL2a promotes B cell development and controls endosomal traffic by cleavage of the invariant chain. *The Journal of Experimental Medicine*, *210*(1), 41–58. http://dx.doi.org/10.1084/jem.20121069.

Schnyder, T., Castello, A., Feest, C., Harwood, N. E., Oellerich, T., Urlaub, H., et al. (2011). B cell receptor-mediated antigen gathering requires ubiquitin ligase cbl and adaptors grb2 and dok-3 to recruit Dynein to the signaling microcluster. *Immunity*, *34*(6), 905–918. http://dx.doi.org/10.1016/j.immuni.2011.06.001.

Stinchcombe, J. C., Majorovits, E., Bossi, G., Fuller, S., & Griffiths, G. M. (2006). Centrosome polarization delivers secretory granules to the immunological synapse. *Nature*, *443*(7110), 462–465. http://dx.doi.org/10.1038/nature05071.

Stinchcombe, J. C., Salio, M., Cerundolo, V., Pende, D., Arico, M., & Griffiths, G. M. (2011). Centriole polarisation to the immunological synapse directs secretion from cytolytic cells of both the innate and adaptive immune systems. *BMC Biology*, *9*(1), 45. http://dx.doi.org/10.1186/1741-7007-9-45.

Suzuki, K., Grigorova, I., Phan, T. G., Kelly, L. M., & Cyster, J. G. (2009). Visualizing B cell capture of cognate antigen from follicular dendritic cells. *The Journal of Experimental Medicine*, *206*(7), 1485–1493. http://dx.doi.org/10.1084/jem.20090209.

Thaunat, O., Granja, A. G., Barral, P., Filby, A., Montaner, B., Collinson, L., et al. (2012). Asymmetric segregation of polarized antigen on B cell division shapes presentation capacity. *Science*, *335*(6067), 475–479. http://dx.doi.org/10.1126/science.1214100.

Tolar, P., Sohn, H. W., & Pierce, S. K. (2008). Viewing the antigen-induced initiation of B-cell activation in living cells. *Immunological Reviews*, *221*, 64–76. http://dx.doi.org/10.1111/j.1600-065X.2008.00583.x.

Treanor, B., Depoil, D., Bruckbauer, A., & Batista, F. D. (2011). Dynamic cortical actin remodeling by ERM proteins controls BCR microcluster organization and integrity. *The Journal of Experimental Medicine*, *208*(5), 1055–1068. http://dx.doi.org/10.1084/jem.20101125.

Treanor, B., Depoil, D., Gonzalez-Granja, A., Barral, P., Weber, M., Dushek, O., et al. (2010). The membrane skeleton controls diffusion dynamics and signaling through the B cell receptor. *Immunity*, *32*(2), 187–199. http://dx.doi.org/10.1016/j.immuni.2009.12.005.

Tribulatti, M. V., Cattaneo, V., Hellman, U., Mucci, J., & Campetella, O. (2009). Galectin-8 provides costimulatory and proliferative signals to T lymphocytes. *Journal of Leukocyte Biology*, *86*(2), 371–380. http://dx.doi.org/10.1189/jlb.0908529.

Tsai, C.-M., Chiu, Y.-K., Hsu, T.-L., Lin, I.-Y., Hsieh, S.-L., & Lin, K.-I. (2008). Galectin-1 promotes immunoglobulin production during plasma cell differentiation. *Journal of Immunology (Baltimore, Md: 1950)*, *181*(7), 4570–4579. Retrieved from, http://www.ncbi.nlm.nih.gov/pubmed/18802059.

Tsai, C.-M., Guan, C.-H., Hsieh, H.-W., Hsu, T.-L., Tu, Z., Wu, K. J., et al. (2011). Galectin-1 and galectin-8 have redundant roles in promoting plasma cell formation. *Journal of Immunology (Baltimore, Md: 1950)*, *187*(4), 1643–1652. http://dx.doi.org/10.4049/jimmunol.1100297.

Vascotto, F., Lankar, D., Faure-André, G., Vargas, P., Diaz, J., Le Roux, D., et al. (2007). The actin-based motor protein myosin II regulates MHC class II trafficking and BCR-driven antigen presentation. *The Journal of Cell Biology*, *176*(7), 1007–1019. http://dx.doi.org/10.1083/jcb.200611147.

Villadangos, J. A., Riese, R. J., Peters, C., Chapman, H. A., & Ploegh, H. L. (1997). Degradation of mouse invariant chain: Roles of cathepsins S and D and the influence of major histocompatibility complex polymorphism. *The Journal of Experimental Medicine*, *186*(4), 549–560. Retrieved from, http://www.pubmedcentral.nih.gov/articlerender.fcgi?artid=2199027&tool=pmcentrez&rendertype=abstract.

Wubbolts, R., Fernandez-Borja, M., Jordens, I., Reits, E., Dusseljee, S., Echeverri, C., et al. (1999). Opposing motor activities of dynein and kinesin determine retention and transport of MHC class II-containing compartments. *Journal of Cell Science*, *112*(Pt 6), 785–795. Retrieved from, http://www.ncbi.nlm.nih.gov/pubmed/10036229.

Wubbolts, R., Fernandez-Borja, M., Oomen, L., Verwoerd, D., Janssen, H., Calafat, J., et al. (1996). Direct vesicular transport of MHC class II molecules from lysosomal structures to the cell surface. *The Journal of Cell Biology*, *135*(3), 611–622. Retrieved from, http://www.pubmedcentral.nih.gov/articlerender.fcgi?artid=2121075&tool=pmcentrez&rendertype=abstract.

Yuseff, M.-I., Lankar, D., & Lennon-Duménil, A.-M. (2009). Dynamics of membrane trafficking downstream of B and T cell receptor engagement: Impact on immune synapses. *Traffic (Copenhagen, Denmark)*, *10*(6), 629–636. http://dx.doi.org/10.1111/j.1600-0854.2009.00913.x.

Yuseff, M.-I., Pierobon, P., Reversat, A., & Lennon-Duménil, A.-M. (2013). How B cells capture, process and present antigens: A crucial role for cell polarity. *Nature Reviews. Immunology*, *13*(7), 475–486. http://dx.doi.org/10.1038/nri3469.

Yuseff, M.-I., Reversat, A., Lankar, D., Diaz, J., Fanget, I., Pierobon, P., et al. (2011). Polarized secretion of lysosomes at the B cell synapse couples antigen extraction to processing and presentation. *Immunity*, *35*(3), 361–374. http://dx.doi.org/10.1016/j.immuni.2011.07.008.

CHAPTER THREE

Force Generation in B-Cell Synapses: Mechanisms Coupling B-Cell Receptor Binding to Antigen Internalization and Affinity Discrimination

Pavel Tolar[1], Katelyn M. Spillane
Division of Immune Cell Biology, National Institute for Medical Research, London, United Kingdom
[1]Corresponding author: e-mail address: ptolar@nimr.mrc.ac.uk

Contents

Abstract

The B-cell receptor (BCR) controls B-cell activation by biochemical signaling and by physical acquisition of antigens from immune synapses with antigen-presenting cells. B cells grab and gather antigens by engaging conserved biomechanical modules for cell spreading, receptor clustering, receptor transport, and generation of pulling forces, which culminate in antigen extraction and endocytosis. The mechanical activity in B-cell synapses follows a pattern of positive and negative feedbacks that regulate the amount of extracted antigen by directly manipulating the dynamics of BCR–antigen bonds. In particular, spreading and clustering increase the association of BCR with antigen, providing amplification and sensitivity, while pulling forces dissociate the BCR from the antigen, testing the quality of antigen binding. The emergent effect of mechanical forces in B-cell

Advances in Immunology, Volume 123
ISSN 0065-2776
http://dx.doi.org/10.1016/B978-0-12-800266-7.00002-9

synapses is ligand discrimination that can be scaled across a range of BCR affinities, provided that the magnitude and timing of the mechanical forces are precisely coordinated with biochemical readouts from the BCR. Such coordination predicts not only novel connections between BCR signaling, endocytosis, and the actomyosin cytoskeleton but also mechanosensitivity of these pathways. The mechanical control of bond formation and separation may be generally beneficial in signaling networks with variable thresholds.

1. INTRODUCTION: WHY DO B CELLS MAKE IMMUNE SYNAPSES?

It has been known for decades that T cells encounter cognate antigens in cell–cell contacts with antigen-presenting cells (APCs). These contacts, known as immune synapses, provide an organized scaffold that mediates binding between the T-cell receptor (TCR) and antigenic peptides bound to MHC molecules for efficient T-cell activation. Recent evidence indicates that B cells also make immune synapses to recognize intact, unprocessed antigens on APCs (Batista, Iber, & Neuberger, 2001; Carrasco & Batista, 2006b; Cyster, 2010). These results were initially surprising, because B cells have been known to respond well to soluble antigens *in vitro*. However, imaging of B-cell activation in intact lymph nodes showed that B-cell responses to soluble antigen are limited to antigens that are smaller than approximately 70 kDa (Gretz, Norbury, Anderson, Proudfoot, & Shaw, 2000; Pape, Catron, Itano, & Jenkins, 2007; Sixt et al., 2005). The majority of antigens are quickly captured by lymph node-resident APCs, such as subcapsular macrophages and a subpopulation of dendritic cells (Carrasco & Batista, 2007; Gonzalez et al., 2010; Junt et al., 2007; Phan, Grigorova, Okada, & Cyster, 2007), which recycle and display the antigens on their surfaces for presentation to B cells (Bergtold, Desai, Gavhane, & Clynes, 2005; Heesters et al., 2013; Qi, Egen, Huang, & Germain, 2006). Binding of the antigens by B-cell receptors (BCRs) leads to the formation of immune synapses and stimulation of the B cells, making the immune synapse a conserved feature of lymphocyte activation.

High-resolution imaging of B-cell interactions with model APCs *in vitro* has revealed that B cells form synapses within several minutes of the initial contact (Batista et al., 2001; Fleire et al., 2006; Tolar, Hanna, Krueger, & Pierce, 2009). Critical to this process is the reorganization of the membrane cytoskeleton (Harwood & Batista, 2011), which begins just seconds after BCR binding to antigen. BCR signaling orchestrates actin polymerization, which pushes the B-cell membrane into contact with the APC and aligns the surfaces of the two cells together (Fleire et al., 2006; Tolar, Hanna, et al., 2009). The BCR and antigen are compacted into microclusters, which initially contain just a few BCRs and quickly grow into larger structures containing hundreds of BCRs and signaling molecules. The clusters are dragged to the center of the synapse and are repeatedly pulled upon by contractions of the actomyosin cytoskeleton, eventually leading to endocytosis of the antigen (Natkanski et al., 2013). BCR signaling initiated from the synapse and the endosomes activates intracellular pathways that orchestrate antigen processing and transcriptional changes in the nucleus.

How does the complex organization of the immune synapse contribute to B-cell function? Given that B cells can respond to soluble antigen, it is not immediately obvious why B-cell activation requires complex synaptic organization. One possible answer is that the cytoskeleton scaffolding in immune synapses builds unique regulatory networks from common signaling modules, which allowed evolution of lymphocyte synapses adapted to specialized sensory and effector functions. For example, helper T cells generate immune synapses that robustly detect and quantify weak signals (Davis et al., 2007), cytotoxic T cells and NK cells make synapses that deliver lytic granules to kill their targets (Griffiths, Tsun, & Stinchcombe, 2010), and regulatory T cells scavenge stimulatory molecules from the surfaces of the APCs (Qureshi et al., 2011). All of these tasks require specific mechanical functions that are provided by dynamic cytoskeletal machinery. B-cell synapses appear to be "endocytic synapses," in that they specialize in the gathering and endocytosis of antigens. Extraction of APC-bound antigens is a physical process that requires mechanical forces, but the generation of forces also serves the more nuanced role of providing a way for B cells to test BCR affinity for antigen (Batista & Neuberger, 2000; Natkanski et al., 2013). Thus, antigen extraction occurs in conjunction with reading out the quality of BCR binding.

To read out the affinity of the BCR–antigen bonds, B cells apply complex signaling pathways to interpret the binding parameters. Previously, much attention has been devoted to the role of the cytoskeleton in regulating these pathways by restricting diffusion of membrane molecules (Harwood & Batista, 2011). Indeed, trapping and corralling membrane molecules regulate lateral interactions and thereby modulate the signaling in the plasma membrane (Sohn, Tolar, & Pierce, 2008; Tolar, Hanna, et al., 2009; Treanor, Depoil, Bruckbauer, & Batista, 2011; Treanor et al., 2010). However, it is difficult to ignore the mechanical activity of the cytoskeleton itself. In the immune synapse, the cytoskeleton generates forces to physically bring receptors to the ligands, cluster them, and then pull them apart. Thus, mechanical activity changes the effective association and dissociation rates of the receptors and can therefore be used as a method to directly test the relationship between binding and signaling. Importantly, the critical outcome of B-cell synapse formation—antigen internalization—is a physical result of BCR–antigen binding and, as such, is difficult to regulate purely biochemically.

The mechanical testing of ligand binding by B cells can be thought of as testing the stickiness of a surface with your finger. When you touch something—whether a clean tile or frosting on a cake—there are binding interactions that occur between the surface and your finger. As you retract your hand, the feedback you receive lets you know how much force is required to detach your finger and also how much surface material remains adhered to it. The touch sensors in our fingertips allow us to respond to this complicated feedback intuitively by adjusting the contact time and the speed of retraction. Similarly, B cells must be able to adjust their mechanical response to adapt to a wide range of BCR-antigen affinities.

We focus in the succeeding text on the molecular mechanisms that control the mechanical testing of antigen binding. These appear to be conserved modules for membrane extension and retraction, receptor clustering, transport, and endocytosis. Finally, we discuss how coordination of these biomechanical processes regulates B-cell affinity discrimination.

2. BIOMECHANICAL MODULES IN B-CELL SYNAPSES

Our current understanding of B-cell synapses suggests that their architecture depends on at least five dynamic cytoskeleton-dependent processes (Fig. 3.1). First, spreading is mediated by protrusion of lamellipodia at cell edges. Second, BCR microclusters form in the initial contact areas and in the lamellipodia. Third, BCR microclusters are transported toward the center of the synapse where they accumulate. Fourth, BCR microclusters are pulled upon and invaginated. And fifth, microclusters are endocytosed, extracting the antigen for processing.

2.1. Spreading

When a B cell encounters an APC, its BCRs bind the presented antigen and B-cell spreading is initiated. This process has been directly linked to proximal BCR signaling, which depends on the BCR-associated tyrosine kinases Lyn and Syk, phosphorylation of the membrane adaptor CD19, and recruitment of cytoplasmic proteins BLNK, PLCγ, and Vav (Depoil et al., 2008; Weber et al., 2008). These molecules form a protein complex associated with the BCR that activates signaling pathways responsible for actin remodeling. It is thought that one of the first steps in spreading is severing of the

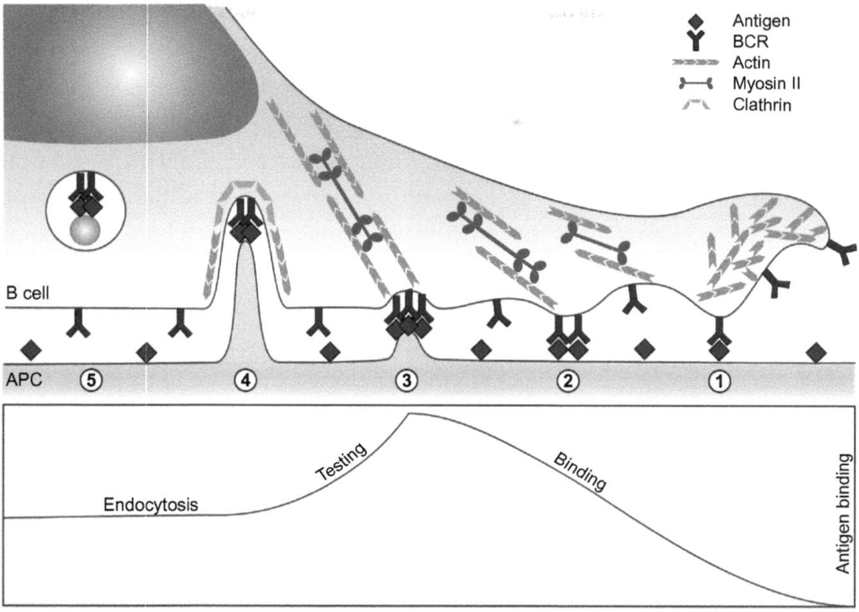

Figure 3.1 *Mechanics of BCR binding at the B-cell synapse.* Schematic depiction of the B-cell synapse (top) and the amount of local antigen binding (bottom). (1) BCR signaling in lamellipodia stimulates actin polymerization, protrusion of the leading edge, and pushing of the B cell membrane into contact with the APC. Binding is limited by the incomplete alignment of the B-cell membrane with the APC. (2) BCR microclusters are transported by actin flow and by myosin IIa contractions toward the center of the contact. Binding is promoted by alignment of the synaptic membranes and by BCR clustering. (3) Myosin IIa pulls on the microclusters and invaginates the B-cell membrane. The forces trigger accelerated dissociation. (4) Formation of clathrin-coated pits (CCPs) generates additional actin polymerization and eventual endocytosis of the antigen. (5) Forces are terminated and endocytosis reads the final number of antigens bound to the BCR. (See the color plate.)

existing cortical actin cytoskeleton by a pathway involving Rap1 and cofilin (Freeman et al., 2011). Actin depolymerization, together with decoupling of the plasma membrane from the actin through dephosphorylation of the cytoskeletal adaptors ezrin and moesin, promotes changes of the plasma membrane shape and BCR binding (Treanor et al., 2011). Activation of Vav stimulates actin polymerization through activation of Rac and the Arp 2/3 complex. Actin polymerization drives extension of the membrane edge and generation of lamellipodia. In B cells lacking components of this pathway, B-cell synapses are smaller and the alignment of the B cell with the presenting membrane is imperfect. As a result, these B cells produce fewer BCR microclusters and less signaling occurs (Arana et al., 2008; Depoil et al., 2008; Weber et al., 2008).

The contact of lamellipodia with the APC is assisted by integrin-mediated adhesion, which stabilizes the synapse and contributes to the overall synaptic architecture. In naive B cells, the integrins LFA-1 and VLA-4 are rapidly activated by BCR signaling and engage their ligands on the APCs (Carrasco & Batista, 2006a; Carrasco, Fleire, Cameron, Dustin, & Batista, 2004). After a short period of actin reorganization and initial membrane alignment, the extension of the lamellipodia proceeds through cycles of protrusions and retractions (Tolar, Hanna, et al., 2009), which resemble the contractile spreading described in mesenchymal cells (Gauthier, Masters, & Sheetz, 2012). The protrusions are driven by actin polymerization at the leading edge of the cell, while the retractions are mediated by the non-muscle myosin IIa pulling from the back of the lamellipodia (Giannone et al., 2007). The pulling by myosin IIa curls the actin and lifts the leading edge of the substrate in a search of antigen. Thus, the lamellipodial activity generates two forces on the BCR. First, actin polymerization leads to actin flow and generates lateral forces that initiate the centripetal movement of the BCRs. These forces are generally small. Second, the intermittent myosin IIa contractions generate strong forces that may underlie the first mechanical testing of BCR binding. This interactive model of spreading of the leading edge is consistent with the view that B-cell spreading is correlated with the strength of BCR binding to antigen. As a result, the area of the synapse correlates well with affinity of the BCR for antigen (Fleire et al., 2006; Liu, Meckel, Tolar, Sohn, & Pierce, 2010).

2.2. Clustering

BCR microclusters form simultaneously with BCR binding to antigen and continue to grow by recruitment of both antigen and BCR molecules. Although multivalency of the antigen likely contributes to BCR microclustering, it is not an absolute requirement (Fleire et al., 2006; Tolar, Hanna, et al., 2009). Instead, BCR microclusters form by a surprisingly complex process that involves several mechanisms (Pierce & Liu, 2010; Tolar, 2011). Initially, membrane curvature at the contact sites can lead to diffusional confinement of antigen-engaged BCRs. The engaged BCRs then become trapped in a tightly packed cluster via oligomerization (Tolar, Hanna, et al., 2009; Tolar & Pierce, 2010; Tolar, Sohn, Liu, & Pierce, 2009). These microclusters are also corralled by the actin cytoskeleton (Treanor et al., 2011), which increases the rebinding rate of dissociated BCRs. Further growth of the microclusters is possible through continuous trapping of diffusing BCRs and also by active movement of BCR clusters centripetally, which merges small microclusters together.

Exactly how BCRs form the initial antigen-induced microclusters is still not completely understood. Superresolution imaging shows that a significant fraction of BCRs in the membranes of naive B cells are organized in higher-order structures (Mattila et al., 2013). These structures, or islands, are areas of the plasma membrane with a radius of about 60 nm that contain 20–100 BCRs along with a number of other proteins. This organization is particularly dramatic for the IgD isotope of the BCR, which is highly expressed on naive follicular B cells. The dynamics of these protein islands are not well understood, although single-molecule tracking shows that most BCR are mobile, indicating that these protein islands must dynamically exchange BCR molecules (Tolar, Hanna, et al., 2009; Treanor et al., 2010). Similarly, the exact function of the nanoscale organization of the BCR is not clear, although it is possible that close proximity of the BCRs and their connections through the underlying cytoskeleton can have cooperative effects on BCR–antigen binding.

2.3. Pulling

Spreading is short-lived in B-cell synapses. After 2–3 min, B-cell synapses enter the contraction phase (Fleire et al., 2006), during which myosin contractility strengthens and retractions of the lamellipodia become greater than the protrusions. Antigen microclusters are dragged with the retracting cell edge as the synaptic area shrinks. After the retraction, lamellipodia protrusions are rare. Myosin contractility continues to pull on antigen microclusters, however, creating local, unsynchronized invaginations throughout the synapse (Natkanski et al., 2013). The invaginations appear to be initiated by actin polymerization directly underneath the BCR microclusters, together with recruitment of myosin IIa. Exactly how myosin IIa pulls the microclusters inward is not clear, however, as myosin is quickly cleared from the immediate proximity of the microcluster, while F-actin remains associated.

Myosin IIa is a hexamer composed of two heavy chains, each of which contains one actin-binding head that generates the power stroke, two essential light chains, and two regulatory light chains (Lecuit, Lenne, & Munro, 2011). Activation of myosin IIa requires serine and threonine phosphorylation of the regulatory light chains, which allow myosin IIa to assemble into minifilaments with the heads pointing in opposite directions. Myosin minifilaments crosslink antiparallel actin cables and the ATP activity of the myosin heads then processively slides the actin cables toward each other, generating contractile force. It is possible that myosin IIa in B-cell synapses invaginates BCR microclusters by contracting actin filaments that are connected to the BCR microclusters on one end and to cytoskeletal structures deeper in the cell on the other end. Alternatively, contraction can occur around BCR microclusters, wrinkling the membrane into the cell. These mechanisms may have implications for the strength of the forces that B cells generate and for the mechanism that distributes the load among BCRs in the microclusters.

Myosin IIa is very dynamic in B-cell synapses and the signaling that regulates its localization and activation is incompletely understood. The phosphorylation of the regulatory light chain seems to be primarily mediated by the kinase Rock1 (Natkanski et al., 2013; Vascotto et al., 2007). The involvement of Rock1 in myosin contractility in B-cell synapses suggests that myosin IIa is activated by the canonical pathway downstream of RhoA. RhoA has been reported to be activated by the BCR (Saci & Carpenter, 2005) and is required for B-cell development (Zhang, Zhou, Lang, & Guo, 2012). However, little is currently known about the pathways that activate RhoA in B-cell synapses and it remains therefore to be established how this pathway connects to BCR signaling.

2.4. Transport

As mentioned above, transport of antigen microclusters toward the center of the synapse is initiated by actin flow and myosin contractility in lamellipodia. After B cells enter the contraction phase, centralization of the antigen in B-cell synapses depends on microtubules. Centripetal alignment of microtubules parallel to the plasma membrane follows the polarization of the microtubule-organizing center (Yuseff, Pierobon, Reversat, & Lennon-Duménil, 2013), and connection of the BCR microclusters to the motor protein dynein then leads to the microclusters tracking the microtubules toward the center of the synapse (Schnyder et al., 2011). The polarization of the microtubule network is also associated with accumulation of endosomes and lysosomes underneath the plasma membrane of the synapse and eventually to fusion of lysosomes with the plasma membrane (Yuseff et al., 2011), which can enzymatically liberate the antigen. The exact role for final localization of the BCR in the center of the synapse is, however, not clear. It is possible that the centripetal transport of antigen microclusters contributes to microcluster growth and therefore to some aspects of positive cooperativity in the microclusters. It is also possible that the center of the synapse is a specific environment that specializes in downmodulation of the signaling. Alternatively, accumulation of BCRs in this area may be in preparation for termination of the contact with the APC, either through final internalization of the antigen or through the lysozyme-mediated enzymatic dissociation of the bonds with the APC.

2.5. Internalization

Most current data are consistent with the view that clathrin-coated pits (CCPs) are the dominant mechanism for antigen internalization in B cells (Busman-Sahay, Drake, Sitaram, Marks, & Drake, 2013; Chaturvedi, Martz, Dorward, Waisberg, & Pierce, 2011; Natkanski et al., 2013; Stoddart et al., 2002; Stoddart, Jackson, & Brodsky, 2005). Genetic disruption of clathrin, or other canonical components of the CCPs, leads to substantial reduction of internalization of both soluble and membrane-presented antigen (Natkanski et al., 2013; Stoddart et al., 2005). Clathrin-independent, lipid raft-, and actin-dependent pathways may also be involved (Stoddart et al., 2005), but the role of these pathways may be only compensatory, as clathrin colocalizes with the great majority of antigen clusters that are internalized (Natkanski et al., 2013).

CCPs initiate their assembly at the plasma membrane by stochastic recruitment of the adaptor AP2 and clathrin (Cocucci, Aguet, Boulant, & Kirchhausen, 2012). The maturation of the CCPs then involves a large number of proteins that are specialized in membrane bending, capture of membrane receptors, actin polymerization, membrane scission, and vesicle uncoating (Doherty & Mcmahon, 2009; Kaksonen, Toret, & Drubin, 2006; Mcmahon & Boucrot, 2011; Taylor, Perrais, & Merrifield, 2011). The energy for bending and invagination of the CCP comes from two sources: proteins that induce or stabilize membrane curvature, such as BAR domain proteins and clathrin itself, and actin polymerization (Stachowiak, Brodsky, & Miller, 2013). Actin polymerization is particularly required when membrane tension is high (Boulant, Kural, Zeeh, Ubelmann, & Kirchhausen, 2011), as appears to be true for B cells, even in the case of internalization of soluble antigen (Malhotra, Kovats, Zhang, & Coggeshall, 2009; Sharma, Orlowski, & Song, 2009). Actin polymerization is also important for extending the endocytic invagination in cases where the cargo exceeds the size of the clathrin-coated vesicle, as may often happen with antigen clusters (Cureton, Massol, Whelan, & Kirchhausen, 2010).

However, although CCPs and the associated actin polymerization are required for antigen internalization, they are not sufficient for extraction of membrane antigens in B-cell synapses. Inhibition of myosin IIa contractility prevents antigen internalization by blocking the invagination of BCR microclusters, which occurs before the formation of the CCP (Natkanski et al., 2013). This resembles maturation of focal adhesions, where the forces mediated by myosin contractility eventually lead to termination of the focal adhesion by CCPs (Caswell, Vadrevu, & Norman, 2009; Kuo, Han, Hsiao, Yates, & Waterman, 2011). Myosin IIa is also required for CCP mediated internalization of cadherins (Levayer, Pelissier-Monier, & Lecuit, 2011). However, the mechanisms by which myosin contractility leads to endocytosis in any of these cases are not fully understood, although it is possible that myosin contractility promotes endocytosis by clustering the receptors, activating force sensors or changing membrane curvature. While these mechanisms may play a role in B-cell synapses, additional myosin IIa-mediated mechanisms could also be necessary, particularly to provide strong forces that are physically required to pull the antigen from the presenting membrane. This possibility is suggested by observations that inhibition of myosin IIa does not inhibit the rate of CCP formation or their colocalization with BCR (Natkanski et al., 2013). Thus, CCPs alone do not seem to be able to generate the forces required for endocytosis of the membrane-attached antigen.

Myosin contractility can promote antigen extraction in two ways. First, pulling by myosin IIa can rupture the bond between the antigen and the antigen-presenting receptors on the APC. Three receptor classes so far have been implicated in the presentation of antigen to B cells: complement receptors, Fc receptors, and lectins (Cyster, 2010). While these bind the antigens with relatively weak affinity, they typically present the antigens in clusters containing a large number of the receptors, suggesting that substantial forces are needed to physically separate the antigen from the APC membrane. Atomic force microscopy (AFM) experiments show that when APC surface-bound immune complexes are pulled, the APC membranes are first extruded into nanotubes before the bonds rupture (Natkanski et al., 2013). This suggests that B cells need to pull BCR microclusters inward up to several hundred nanometers to separate antigen from the APC. In this second case, the antigen does not need to separate from the presenting receptors, but instead, B cells can directly pinch off the invaginated membrane nanotube. This mechanism resembles transendocytosis that has been described in various receptor–ligand systems, such as EphB/ephrin-B (Pitulescu & Adams, 2010) and CTLA-4/CD86 (Qureshi et al., 2011). In *in vitro* experiments, 60–90% of the endocytosed antigen colocalized with a vesicle of the presenting membrane and a similar proportion of antigen acquired by B cells *in vivo* colocalized with transmembrane proteins derived from follicular dendritic cells (Suzuki, Grigorova, Phan, Kelly, & Cyster, 2009). Thus, both separation of the antigen from the APC and transendocytosis are likely to occur, although the advantage of transendocytosis is that it does not depend on weak attachment of the antigen to the APCs.

Once BCR microclusters have been invaginated, they are internalized by apparently stochastic formation of CCPs. In contrast to the extensive characterization of the BCR signaling components, relatively little is known about how the BCR binds to the CCPs. It has been proposed that the tyrosine residues of the BCR's Igα and Igβ bind to the cargo-binding domain of AP-2, the essential adaptor of CCPs (Busman-Sahay et al., 2013; Gazumyan, Reichlin, & Nussenzweig, 2006; Hou et al., 2006; Jang, Machtaler, & Matsuuchi, 2010). Paradoxically to their potential role in endocytosis, the Igα/Igβ tyrosines become rapidly phosphorylated after antigen stimulation, which is predicted to inhibit binding to AP-2. It has been therefore proposed that dephosphorylation of Igα/Igβ is necessary for internalization (Hou et al., 2006). Alternatively, ubiquitination of Igβ and Syk could recruit the BCR to CCPs (Jacob, Todd, Sampson, & Puré, 2008; Sohn, Gu, & Pierce, 2003; Zhang et al., 2007).

Although the physical connection between the BCR and CCPs is not completely understood, it appears that BCR signaling is connected to clathrin dynamics. The numbers of CCPs and their lifetimes increase in B-cell synapses (Natkanski et al., 2013). Notably, the clathrin heavy chain becomes phosphorylated by the Src-family kinase Lyn after antigen stimulation (Stoddart et al., 2002), and Lyn has also been implicated in the regulation of the rate of antigen-triggered endocytosis (Dykstra, Longnecker, & Pierce, 2001; Ma et al., 2001). Although the exact role that clathrin phosphorylation plays in B-cell synapses is not clear, phosphorylation of clathrin has been shown to promote clathrin recruitment to endocytic sites and clathrin-associated actin polymerization (Bonazzi et al., 2011).

While internalization removes the BCR from the plasma membrane, there is accumulating evidence that internalization does not extinguish all BCR signaling (Chaturvedi, Dorward, & Pierce, 2008; Chaturvedi et al., 2011). The signaling from endosomal compartments contributes to B-cell activation and is required for trafficking of the antigen to late endosomes and their fusion with lysosomes, which triggers antigen processing and loading onto MHC II (Brown, Li, Cheng, & Song, 1999). The nature of signals regulating BCR trafficking is poorly understood, but they have profound importance for the outcome of B-cell activation. For example, altered BCR signaling in anergic B cells prevents antigen trafficking to late endosomes, and as a result, BCR accumulates in recycling endosomes (Blery, Tze, Miosge, Jun, & Goodnow, 2006). This prevents antigen presentation and recognition of the antigens by endosomal Toll-like receptors (O'Neill et al., 2009).

3. ANTIGEN BINDING IN B-CELL SYNAPSES

In contrast to the binding of soluble antibodies and antigen, synaptic binding is influenced by the fact that the BCR and antigens are anchored to two opposing membranes in a dynamic cell–cell contact. The contribution of the membranes to the binding can be broken down into a number of factors, such as the entropic effects of the molecular orientation and flexibility of the proteins, the distribution of the proteins on the cellular surfaces, the intermembrane distance, membrane roughness, and the forces that the bonds experience (Chen & Zhu, 2013; Dustin, Bromley, Davis, & Zhu, 2001; Hu, Lipowsky, & Weikl, 2013; Wu, Vendome, Shapiro, Ben-Shaul, & Honig, 2011). Although no direct measurements of 2D binding for the BCR are currently available, measurements of TCR binding to pMHC in T-cell synapses have been made and attest to the notion that 2D binding is difficult to predict purely from soluble 3D binding constants and actually can vary between different regions of a single synapse (Huang et al., 2010; Huppa et al., 2010).

The 2D binding measurements for the TCR show that the on-rates are generally faster in 2D than what predictions from 3D data would suggest (Huang et al., 2010; Huppa et al., 2010). A fast on-rate is greatly dependent on the organization of the receptors (and presumably ligands) in the membrane, as disruption of the membrane domain organization reduces the 2D on-rates. This is consistent with the idea that clustering of receptors can provide positive cooperativity for the on-rates as binding of a single receptor aligns the whole cluster into the ideal proximity for binding of antigens on the opposing surface (Dustin et al., 2001; Huang et al., 2010). Our emerging understanding of the complex membrane organization of the BCR (Mattila et al., 2013; Yang & Reth, 2010) and of the antigen-induced clustering of the BCR (Depoil et al., 2008; Tolar, Hanna, et al., 2009) suggests that these effects are likely important for BCR–antigen binding as well. The relevance of membrane involvement in BCR–antigen binding is illustrated by recent studies of broadly neutralizing antibodies, in which virus-specific affinity-matured antibodies were reverted to their germline versions (Lingwood et al., 2012). The authors found that the germline antibodies bind very poorly in solution, even in the case of multivalent virions. However, when expressed as BCRs on B-cell surfaces, these antibodies do bind the virus, demonstrating the sensitivity of the multivalent interactions in the context of the plasma membrane. In immune synapses, the overall effective binding can be further improved by the signaling-dependent spreading of the B-cell membrane against the APC. Spreading aligns the membranes of the interacting cells and also increases the area of contact, providing a positive feedback loop for BCR binding (Fig. 3.1).

In addition to the measurements of 2D on-rates, studies on T-cell synapses have also revealed much information regarding the lifetimes of receptor–ligand bonds. In contrast to the increased on-rates, the lifetime of TCR-pMHC bonds in T-cell synapses is decreased compared to that in solution (Axmann, Huppa, Davis, & Schutz, 2012; Huang et al., 2010; Huppa et al., 2010), and it is generally thought that this increased off-rate is a consequence of mechanical forces on the bonds (Bao et al., 2010). Forces can originate not only from large-scale movement of the two interacting cells before they come to a complete stop but also from local, microscopic effects. First, passive forces are produced by membrane tension that results from the bending of the membranes, which is necessary to bring the two opposing membranes together for the receptor to reach the ligand (Allard, Dushek, Coombs, & van der Merwe, 2012). These forces may dominate in areas of the synapse, where spreading has not yet established alignment of the plasma membrane with the APC. Second, as discussed in the preceding text, forces may be actively generated by the cellular cytoskeletal machinery (Husson, Chemin, Bohineust, Hivroz, & Henry, 2011). These strong active forces are generated in B-cell synapses by actomyosin contraction activated by BCR signaling (Natkanski et al., 2013) and thus represent a negative feedback for BCR–antigen binding (Fig. 3.1).

There is a considerable level of understanding of how forces accelerate the off-rate of a bond. For a simplified, theoretical case, the half-life of an intermolecular bond under force, t_{F}, can be expressed as function of the force, F:

$$t_{\mathrm{F}} = t_0 \mathrm{e}^{-(x_{\mathrm{b}} F / k_{\mathrm{B}} T)},$$

where t_0 is the half-life in the absence of external force, k_{B} is the Boltzmann constant, T is absolute temperature, and x_{b} is the distance to the energy barrier at which the bond dissociates (Bell, 1978). Experimentally, the effects of forces on the half-lives of molecular bonds have been directly observed using single-molecule techniques, such as optical tweezers and AFM (Brenner,

Zhou, & Ha, 2011; Greenleaf, Woodside, & Block, 2007). These techniques allow the application of force to a single bond and measurement of the corresponding dissociation time. Alternatively, it is possible to move one of the interacting molecules relative to the other with a constant speed, which leads to a steady increase of the force until the bond ruptures. This experimental setting determines how the rupture forces change with the bond loading rate. Mathematically, the results of these two approaches can be interconverted (Dudko, Hummer, & Szabo, 2008; Pierse & Dudko, 2013).

Most experimental force measurements of antibody–antigen bonds agree with the simple model described above (Marshall et al., 2003; Morfill et al., 2008) (Katletz, Stroh, Rankl, Titulaer, & Hinterdorfer, 2010; Morfill et al., 2007). These bonds, whose half-lives decay exponentially with increasing force, have been termed slip bonds. For slip bonds, the rate of decrease in half-life with force is defined by the parameter x_b. This parameter reflects the molecular properties of the binding interface and can be extracted from the single-molecule pulling experiments. For a typical antigen–antibody bond, x_b is on the order of 1 nm, meaning that an antibody with a zero-force half-life of 10 s will have a half-life of less than 100 ms at 20 pN. Such forces can be expected in B-cell synapses, indicating that it is within the strength of B cells to dramatically accelerate dissociation of the BCR from antigen.

However, half-lives of molecular bonds do not necessary need to decrease exponentially with force and sometimes show even a counterintuitive catch bond behavior—within a certain force range, half-lives increase with increasing forces (Marshall et al., 2003; Thomas, 2008; Yakovenko et al., 2008). A general explanation is that forces can induce changes in the interacting proteins, which results in various forces versus half-life scenarios (Kim, Zhang, Zhang, & Springer, 2010; Le Trong et al., 2010; Rakshit, Zhang, Manibog, Shafraz, & Sivasankar, 2012; Suzuki & Dudko, 2011). Interestingly, and in contrast to soluble antibodies, the membrane BCR shows a complex relationship between half-life and force. The half-life of a BCR binding to antigen drops faster than that of free antibody under small forces, but for forces beyond approximately 20 pN, the BCR switches to a mode where half-live declines less steeply (Natkanski et al., 2013). One possibility is that the BCR is binding the antigen initially with an unfavorable geometry, allowing faster dissociation under small forces. Higher forces may extend or reorient the BCR on the cell surfaces and result in a better resistance to force. This scenario suggests a mechanosensitivity in the BCR, but more detailed measurements of this behavior will be required to understand its basis and its role in regulation of antigen binding (Friddle, Noy, & De Yoreo, 2012).

These data illustrate that the plasma membrane and the underlying cytoskeleton can regulate binding of antigen in immune synapses. In addition, the 2D environment can also regulate BCR signaling. For example, plasma membrane lipids and the submembrane cytoskeleton regulate BCR signaling by controlling lateral interactions of the BCR with signaling molecules (Mattila et al., 2013; Sohn et al., 2008; Treanor et al., 2010, 2011). Intriguingly, external forces have been proposed as a direct activator of antigen receptors (Kim et al., 2012; Ma & Finkel, 2010). In B cells, binding of membrane antigens leads to exposure of the membrane proximal domain of the immunoglobulin, which promotes clustering and intracellular signaling (Tolar, Hanna, et al., 2009). However, direct demonstration of force-induced conformational changes in antigen receptors remains to be shown. It can be expected, though, that some mechanosensitivity is required to coordinate antigen binding with cytoskeletal feedback. In addition to membrane receptors, several signaling adaptors are known to be activated by stretching and the resulting force-induced signaling is important for cells to correctly respond to the mechanical properties of their environment (Gomez, McLachlan, & Yap, 2011; Lecuit et al., 2011). In this context, a recent report shows that the BCR induces stronger proximal signaling on antigenic substrates with high stiffness then with low stiffness (Wan et al., 2013), suggesting that BCR signaling is mechanosensitive, and raising the possibility that B-cell responses could be regulated by the stiffness of the APCs.

4. AFFINITY DISCRIMINATION IN B-CELL SYNAPSES

An effective B-cell response requires the production of antibodies that not only are specific but also have a sufficiently high affinity for the invading pathogen. Because most of the specificities in the naive B-cell repertoire have low affinities, antibody responses depend on selective expansion and differentiation of high-affinity B cells over their low-affinity counterparts (Chan & Brink, 2012). Early in the immune response, high-affinity binding to the BCR triggers preferential expansion and differentiation of B cells into plasma cells, which provides immediate antibody production and early protection (Benson, Erickson, Gleeson, & Noelle, 2007; Paus et al., 2006). Low-affinity B cells enter the germinal center reaction, where the antibody genes undergo somatic mutagenesis and clones containing mutations improving the affinity for antigen are selectively expanded (Victora & Nussenzweig, 2012). These affinity-matured clones exit the germinal center and generate high-affinity plasma cells (Dal Porto, Haberman, Kelsoe, & Shlomchik, 2002; Phan et al., 2006). The germinal centers also generate memory B cells, which have, intriguingly, somewhat lower affinity and broader specificity (Tarlinton & Good-Jacobson, 2013). Exactly how B cells make all of these decisions is not completely understood, but it is thought that B-cell affinity discrimination underlies many of these choices in both naive and antigen-experienced cells.

What is the range of affinities that B cells can distinguish? Experiments with naive B cells indicate that B cells can discriminate differences in binding affinity of monovalent antigens over a range of 1 μM to 100 pM (Batista & Neuberger, 1998, 2000). For multivalent antigens, the lower detection limit is likely even lower, in the hundreds of μM range (Schmidt et al., 2013; Wedemayer, Patten, Wang, Schultz, & Stevens, 1997). To determine the affinity of binding over a range spanning many orders of magnitude, B cells cannot simply read out the equilibrium occupancy of the BCR by antigen. A linear readout would be severely limited in sensitivity in the low-affinity range, while it would suffer from saturation in the high-affinity range. Yet, B cells can distinguish small affinity changes. For example, individual somatic mutations that are positively selected in the germinal center increase binding by as little as twofold (Kocks & Rajewsky, 1988). It is thus assumed that a complex regulatory network is in play to capture differences in binding along this affinity range.

The ability of B cells to discriminate higher and higher affinity eventually curtails the ultimate affinity that B-cell responses can generate. Few antibodies mature beyond the 100 pM ceiling. Yet this ceiling is not a physical inability of the antibody structure to bind with even higher affinity, as antibodies have been further matured *in vitro* and can reach the femtomolar range (Boder, Midelfort, & Wittrup, 2000; Steidl, Ratsch, Brocks, Dürr, & Thomassen-Wolf, 2008). The problem is that high-affinity interactions that associate faster and dissociate slower than the rates of the signaling and endocytic steps in B cells make discrimination and selection difficult. Even for moderate affinity B-cell clones, the discrimination is complicated by the fact that antibodies are bivalent and most antigens are multivalent. The avidity of the multivalent interactions quickly reaches the theoretical ceiling of affinity discrimination. In addition, the avidity effect is greater for low-affinity antibodies than for high-affinity antibodies, effectively masking the true affinity (Oda, 2004). Still, B-cell selection *in vivo* is not impeded by the multivalency of pathogens.

How, then, do B cells discriminate between different affinities? It has been shown that certain aspects of B-cell activation are sensitive to affinity, while others are not. For example, higher-affinity antigens induce more phosphorylation of Syk, higher levels of intracellular calcium, and better differentiation into plasma cells than lower-affinity antigens, while the phosphorylation of Lyn or upregulation of MHC II is induced equally well by both low- and high-affinity antigens (Kouskoff et al., 1998). These findings could be explained by the concept of kinetic proofreading (McKeithan, 1995), which proposes that signaling steps that occur on a timescale shorter than the half-life of the antigen-BCR complex will be triggered efficiently, whereas signaling steps that take longer will be triggered poorly. This scenario could work well for affinity discrimination of low-affinity antigens, where the half-lives of the BCR–antigen bonds are similar to the timescales of the signaling reactions. For BCR–antigen complexes that are stable for more than a few seconds, however, this mechanism alone cannot explain B-cell affinity discrimination.

Importantly, experiments following B-cell selection *in vivo* point out that the preferential expansion of high-affinity B-cell clones is a consequence of B-cell clonal competition and must therefore have a B cell-extrinsic component (Dal Porto et al., 2002; Shih, Meffre, Roederer, & Nussenzweig, 2002; Shih, Roederer, & Nussenzweig, 2002). The extrinsic factor that drives B-cell clonal competition, particularly for selection in the germinal centers, has been identified as T-cell help (Allen, Okada, Tang, & Cyster, 2007; Victora et al., 2010). This suggests that the internalization and presentation of antigen are the most important outcome of the B cell-intrinsic affinity discrimination and must therefore quantitatively reflect the affinity of the BCR for antigen. Indeed, high-affinity B cells do acquire more antigen than low-affinity B cells *in vivo* (Schwickert et al., 2011; Suzuki et al., 2009). Thus, in addition to controlling the specificity of the response, B-cell–T-cell collaboration is a quantitative mechanism for B-cell affinity discrimination. Exactly how T cells allot more help to high-affinity B cells than to low-affinity B cells is not clear, but this mechanism is consistent with evidence showing that T cells make synapses preferentially with targets providing higher stimulus to their TCR (Depoil et al., 2005). Thus, although there are other factors that control how much antigen each B-cell clone can see, such as epitope availability and competition with soluble antibodies, T-cell help is the most important factor for B-cell clonal selection (Victora & Nussenzweig, 2012).

Given the importance of antigen presentation for T cell-mediated selection of high-affinity B-cell clones, it is puzzling that internalization of soluble antigens by B cells *in vitro* is not always a very sensitive readout of affinity. It has been shown that B cells can internalize more high-affinity antigen than low-affinity antigen, provided that the antigen is monovalent (Batista & Neuberger, 1998). However, internalization of multivalent antigens is not very dependent on affinity (Batista & Neuberger, 2000; Kouskoff et al., 1998; Natkanski et al., 2013). These results suggest that endocytosis of soluble antigens simply reflects the equilibrium occupancy of the BCR and is therefore prone to the confounding effects of avidity. A possible reconciliation of these observations has been offered by studies of B-cell immune synapses (Batista et al., 2001; Batista & Neuberger, 2000). These and later studies found that B cells internalize and present antigens in a highly affinity-dependent manner when the antigens are presented on a substrate (Batista & Neuberger, 2000; Fleire et al., 2006; Natkanski et al., 2013). Detailed imaging of the gathering of the antigens in B-cell synapses showed that high-affinity antigens triggered stronger B-cell spreading and therefore access to more antigens than low-affinity antigens (Fleire et al., 2006). Analysis of individual microclusters in B-cell synapses showed that high-affinity antigens promote faster growth of the microclusters and stronger recruitment of signaling molecules to the BCR (Liu et al., 2010). Finally, the extraction and endocytosis of high-affinity antigens occurred with greater efficiency (Natkanski et al., 2013). Importantly, antigen gathering and endocytosis of membrane antigens are dependent on intact BCR signaling and the actomyosin cytoskeleton. These results suggest that the affinity-dependent endocytosis of antigens from B-cell synapses is an active and adaptive process, which requires mechanical activity.

4.1. Forces in affinity discrimination

How can the mechanical activity of the cytoskeleton in B-cells synapses promote affinity discrimination? Current data indicate that affinity discrimination occurs in individual microclusters (Liu et al., 2010). To illustrate the effects of forces on affinity discrimination in individual microclusters, we simulated binding of membrane BCR to a cluster of multivalent antigens with a range of affinities using a stochastic model (Fig. 3.2). In the absence of external forces, BCR binding quickly reaches saturation for all but the lowest affinity (Fig. 3.2A, left panel). This is perhaps viewed more clearly by calculating the number of antigen molecules bound at the end of each simulation, which in this case is the number of antigens that are internalized (Fig. 3.2A, right panel). In this scenario, all but the lowest affinity BCR

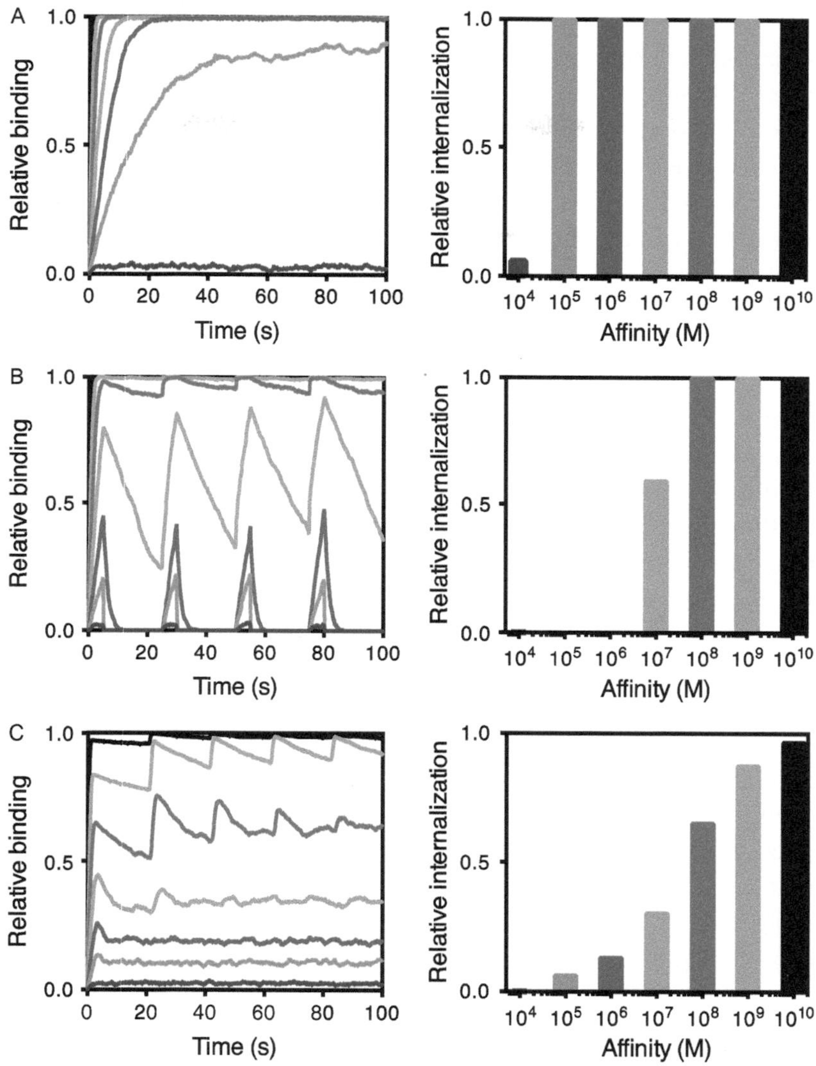

Figure 3.2 *Effect of force on BCR–antigen binding and antigen internalization.* Binding of BCR to antigen was simulated by a stochastic membrane diffusion model using bivalent BCR and a single cluster of immobile multivalent antigens. Simulations were performed with 2D binding affinities that approximately correspond to the 3D affinities indicated in the right panels. The results of the simulation were averaged over 100 runs. (A) Left: Binding as a function of time in the absence of force. Right: Amount of antigen internalized at the end of the simulation. (B) Left: Binding under application of a periodic force. Each force cycle includes a 5-s period with zero force followed by a 20-s period of 50 pN of force. Right: Amount of antigen internalized following the first 25 s force cycle. (C) Left: Binding with application of adaptive forces. Forces were triggered after a threshold number of bonds were formed and their magnitude was dependent on the

(Continued)

internalize antigen efficiently. In contrast, application of a 50 pN periodic force results in discernible association and dissociation phases of the binding, similar to a kinetic binding experiment (Fig. 3.2B, left panel). For simplicity, it is possible to view initial BCR–antigen binding and formation of BCR microclusters as the association phase, when bond numbers per microcluster are increasing (Fig. 3.1, bottom). The bond accumulation rate depends somewhat on affinity, but primarily on the local 2D on-rate. The dissociation phase begins as the contractile machinery triggers a negative feedback and applies a pulling force on the microcluster. The pulling effectively turns off association and starts accelerated dissociation of the bonds. The rate of dissociation is dependent on the off-rate, on the mechanical resistance and on the effective force per bond (the force applied on the microcluster divided by the number of bonds per microcluster). In this simulation, internalization was calculated as the number of antigen molecules bound at the end of the first 20 s window of force application. The results suggest that a periodic application of force inhibits internalization by low-affinity BCRs while permitting internalization by high-affinity BCRs (Fig. 3.2B, right panel) and can therefore provide basic affinity discrimination.

Affinity discrimination in the model improves when the magnitude and timing of the force are adapted to the readout from the BCR (Fig. 3.2C). In this case, better BCR binding triggers stronger forces that occur more quickly following antigen binding. The forces are applied for 20 s or until all bonds in the microcluster rupture, after which a new cycle begins. The use of adaptive forces leads to better separation of the BCR binding curves over a broad range of the affinities, although the individual unsynchronized binding cycles are largely averaged out (Fig. 3.2C, left panel). To simulate endocytosis by CCPs, internalization was calculated as the number of antigen molecules bound whenever the bonds in the microclusters sustained the applied force for the 20 s period, the approximate assembly time for the CCP. The results show that under these conditions,

Figure 3.2—Cont'd number of BCRs in the cluster. Forces were applied for periods of 20 s or until all bonds ruptured. In the case of low-affinity interactions, the bonds rupture frequently, leading to a higher frequency of binding attempts and an overall lower amount of binding. For high-affinity interactions, the bonds withstand the forces, leading to lower frequency of binding events and an overall higher amount of binding. Note that high-frequency information is lost due to averaging of the simulation runs. Right: Amount of antigen internalized in the first instance that the bonds survive a 20 s force application. In all cases, forces on the cluster were divided equally onto all bonds. Dissociation under force was determined using the formula indicated in the text, with $x_b = 1$ nm for all affinities. (See the color plate.)

adaptive forces improve affinity-dependent antigen internalization over a broad range of affinities (Fig. 3.2C, right panel).

This simple model illustrates that mechanical forces can improve affinity discrimination of the BCR and suggests that coordination of the forces and BCR signaling is a critical aspect of the organization of B-cell synapses. Although we do not yet fully understand the molecular connections between the biochemical and the mechanical pathways in B-cell synapses, available observations allude to their existence. Quantification of the acto-myosin localization in B-cell synapses indicates that the amount of myosin IIa and F-actin correlates with the size of antigen microclusters (Natkanski et al., 2013). Since the direct activation of myosin IIa by the BCR has been described (Vascotto et al., 2007), it is conceivable that BCR signaling controls the timing and intensity of myosin contractility. Measurements also showed that complete rupture of BCR–antigen bonds in microclusters leads to rapid depolymerization of the F-actin associated with the microcluster (Natkanski et al., 2013). Thus, either the release of the tension or abrogation of the signaling in the BCR microcluster upon bond rupture is detected by the cytoskeletal elements.

As a result of the adaptive forces, low-affinity binding leads to a higher binding cycle frequency and relatively little cluster internalization, while higher-affinity binding leads to lower-cycle frequency with more clusters being endocytosed. The process is amplified by spreading, which gives the B cells access to a greater number of antigen molecules and the statistical power to perform the pulling experiment hundreds of times in each synapse. This cyclical activity can be viewed as a result of the mechanical positive and negative feedbacks on the BCR binding. A combination of positive and negative feedbacks is a common mechanism in biological oscillators and has been shown to promote network robustness in the face of a broad range of inputs (Tsai et al., 2008).

The combination of positive and negative feedbacks is not limited to the mechanical activity in B-cell synapses. However, the mechanical control of bond formation has certain unique advantages. In a molecular analogy of the finger touch test, it has been experimentally shown that mechanical forces can provide very sharp discrimination between weak and strong binding (Albrecht et al., 2003). Theoretical work also shows that force-mediated dissociation of bonds under increasing load has a narrower distribution than spontaneous dissociation, enhancing the discrimination (Klotzsch & Schutz, 2013). In addition, force resistance can be an independent parameter of the binding, which can help to define the low and high affinity. Finally, because mechanical testing of bond strength works under accelerated association and dissociation, it is fast and could be especially useful for affinity discrimination of bonds that are long-lived in the absence of force. This is important because B-cell synapses *in vivo* last only 5–10 min, suggesting that B cells have limited time for affinity discrimination (Schwickert et al., 2007).

The exact magnitude of the forces on individual BCR–antigen bonds in B-cell synapses awaits experimental measurements. Predictions of the model suggest that the maximum strength of the forces determines the ultimate ceiling for affinity discrimination. Apart from the movement generated by myosin contractility, the forces on the BCR–antigen bonds depend on two additional factors: the linkage of the actomyosin to the BCR and the viscoelastic properties of the APCs. It is likely that the BCR connects to the contractile filaments indirectly, by a mechanism that allows sharing of the force between parallel BCR bonds on the microcluster. Load sharing is important for extraction of low and moderate affinity antigens, because individual bonds do not withstand the pulling forces required for antigen extraction (Sulchek, Friddle, & Noy, 2006). This is supported by findings that larger antigen microclusters withstand extraction forces for longer periods of time (Natkanski et al., 2013).

From the view of the APC, B-cell pulling on the antigen leads first to elastic dissociation of the presenting receptors from the APC cytoskeleton and then to outward viscoelastic pulling of the membrane (Evans & Calderwood, 2007; Nowak & Chou, 2010). This suggests that the strength of binding of the antigen-presenting receptors to the actin cytoskeleton of the APC may provide regulation of B-cell antigen extraction. Once separated from the cytoskeleton, the APC's plasma membrane is pulled into a membrane nanotube. Calculations and force measurements show that during pulling of nanotubes, forces initially increase and then collapse into a plateau phase (Derényi, Jülicher, & Prost, 2002; Krieg, Helenius, Heisenberg, & Müller, 2008). The initial peak of the force depends on the size of the membrane patch, suggesting that larger antigen clusters may require stronger forces for internalization. The forces during the plateau phase depend primarily on the speed of pulling. Thus, the forces on the BCR–antigen bonds must be dynamic in B-cell synapses and can be regulated also by the APCs.

4.2. Implications for affinity maturation

It is well established that antibody affinity maturation improves binding to antigen both by increasing the on-rates and by decreasing the off-rates (Manivel, Sahoo, Salunke, & Rao, 2000; Sagawa, Oda, Ishimura, Furukawa, & Azuma, 2003; Wedemayer et al., 1997). The affinity discrimination model presented above is dependent on both the association and the dissociation phase and therefore predicts selection for on-rates, for off-rates, and potentially also for mechanical resistance. Although it is difficult to generalize for all antibodies, most structural data and biophysical measurements indicate that the underlying molecular basis for affinity maturation is reduced conformational heterogeneity in the antibody's complementarity-determining regions (CDR) (Jimenez, Salazar, Yin, Joo, & Romesberg, 2004; Manivel et al., 2000; Schmidt et al., 2013; Thorpe & Brooks, 2007; Zimmermann et al., 2006). The germline repertoire of antibodies contains highly flexible CDR loops, which allow each B-cell clone to recognize a number of antigens (Willis, Briney, DeLuca, Crowe, & Meiler, 2013). Binding of germline antibodies to a particular antigen thus requires conformational changes in the CDRs, which carries an entropic penalty to the binding, resulting in a low-affinity interaction. Mutations that are selected in germinal centers stabilize CDR loops in a configuration that has improved complementarity with the antigen. Apart from reducing the off-rate, this preconfiguration of the antibody removes the entropic penalty, dramatically enhancing the on-rate. As a result, affinity-matured antibodies also have better specificity for the selecting antigen. A different scenario occurs in chronic infections, as in the case of HIV infection. Reactivation of B cells during viral antigenic escape sometimes leads to the development of broadly neutralizing antibodies, which are specific to a number of variants of the virus. These antibodies are unusual in that they accumulate a large number of mutations, even in regions outside of the CDRs. These distal mutations are critical for binding the virus, although surprisingly, they primarily increase the flexibility of the CDRs, allowing more atomic contacts with the antigen and thus reducing the off-rates (Klein et al., 2013). These results suggest that there is a trade-off between flexibility and specificity of antibodies and that affinity selection may need to balance these factors.

In contrast to the effects of affinity maturation on the kinetic and thermodynamic properties of antibody–antigen bonds, their mechanical stability has not been systematically analyzed. As mentioned previously, antibody–antigen bonds are slip bonds and their mechanical properties can be measured by single-molecule force experiments (Ros et al., 1998). It is known that for affinity-matured antibodies, the exponential decay of bond half-lives with force can be characterized by single x_b value (Katletz et al., 2010; Moreno et al., 2011; Morfill et al., 2007; Schwesinger et al., 2000). Some data suggest a negative correlation between the binding energy and x_b of antibodies (Schwesinger et al., 2000). However, so far, there are no studies comparing germline antibodies to their affinity-matured versions, and thus, the effects of the mutations acquired during affinity maturation *in vivo* on the mechanical stability remain unknown.

Putting aside the possibility that mechanically stable antibody bonds could be beneficial for phagocytosis and cell-mediated cytotoxicity, it could be argued that, ultimately, antibodies need to bind to antigen as soluble molecules and it would not be useful to select for antibodies that bind well only under forces in the 2D environment of the B-cell synapse. However, it is unlikely that such a selection would occur because, for example, the flexibility of the antibody hinge region and the random orientation of the antigens on the APCs would likely mitigate systematic differences between the 2D environment and solution. Nevertheless, differences in mechanical stability of BCR–antigen bonds between different B-cell clones could be an important biophysical factor in their activation, and direct measurements of the mechanical properties of physiologically relevant antibodies may be therefore pertinent to our understanding of the clonal composition of B-cell responses *in vivo*.

5. CONCLUSIONS AND PERSPECTIVE

Recent advancements in live-cell imaging have revealed that signaling networks are organized into intricate molecular assemblies. This is particularly striking in immune synapses. Mechanical actuation and sensing are inherently required for construction and operation of these complex structures, and it is becoming clear that integration of mechanobiology into our analysis of signaling will yield a better quantitative understanding of cellular responses. Understanding the causal relationship between biochemical signaling and mechanical activity in the immune synapse will require development of new tools beyond state-of-the-art imaging, such as quantitative real-time reporters, inducible and spatially controlled genetics, and novel methods of mechanical manipulation of cells. With these tools, it will be possible to address questions about the role of mechanobiology in B-cell responses, including specialization of B-cell subsets and their physiological APCs. Our understanding of these issues may lead to new ways of controlling the affinity and specificity of antibodies produced in vaccination and new methods in the development of high-affinity therapeutic antibodies.

ACKNOWLEDGMENT

The authors' research has been supported by the UK Medical Research Council (Unit Programme number U117597138).

REFERENCES

Albrecht, C., Blank, K., Lalic-Mülthaler, M., Hirler, S., Mai, T., Gilbert, I., et al. (2003). DNA: A programmable force sensor. *Science, 301*, 367–370.

Allard, J. F., Dushek, O., Coombs, D., & van der Merwe, P. A. (2012). Mechanical modulation of receptor-ligand interactions at cell-cell interfaces. *Biophysical Journal, 102*, 1265–1273.

Allen, C. D. C., Okada, T., Tang, H. L., & Cyster, J. G. (2007). Imaging of germinal center selection events during affinity maturation. *Science, 315*, 528–531.

Arana, E., Vehlow, A., Harwood, N. E., Vigorito, E., Henderson, R., Turner, M., et al. (2008). Activation of the small GTPase Rac2 via the B cell receptor regulates B cell adhesion and immunological-synapse formation. *Immunity, 28*, 88–99.

Axmann, M., Huppa, J. B., Davis, M. M., & Schutz, G. J. (2012). Determination of Interaction Kinetics between the T Cell Receptor and Peptide-Loaded MHC Class II via Single-Molecule Diffusion Measurements. *Biophysical Journal, 103*, L17–L19.

Bao, G., Kamm, R. D., Thomas, W., Hwang, W., Fletcher, D. A., Grodzinsky, A. J., et al. (2010). Molecular biomechanics: The molecular basis of how forces regulate cellular function. *Molecular & Cellular Biomechanics, 3*, 91–105.

Batista, F. D., Iber, D., & Neuberger, M. S. (2001). B cells acquire antigen from target cells after synapse formation. *Nature, 411*, 489–494.

Batista, F. D., & Neuberger, M. S. (1998). Affinity dependence of the B cell response to antigen: A threshold, a ceiling, and the importance of off-rate. *Immunity, 8*, 751–759.

Batista, F. D., & Neuberger, M. S. (2000). B cells extract and present immobilized antigen: Implications for affinity discrimination. *The EMBO Journal, 19*, 513–520.

Bell, G. I. (1978). Models for the specific adhesion of cells to cells. *Science, 200*, 618–627.

Benson, M. J., Erickson, L. D., Gleeson, M. W., & Noelle, R. J. (2007). Affinity of antigen encounter and other early B-cell signals determine B-cell fate. *Current Opinion in Immunology, 19*, 275–280.

Bergtold, A., Desai, D., Gavhane, A., & Clynes, R. (2005). Cell surface recycling of internalized antigen permits dendritic cell priming of B cells. *Immunity, 23*, 503.

Blery, M., Tze, L., Miosge, L. A., Jun, J. E., & Goodnow, C. C. (2006). Essential role of membrane cholesterol in accelerated BCR internalization and uncoupling from NF-kappa B in B cell clonal anergy. *The Journal of Experimental Medicine, 203*, 1773–1783.

Boder, E. T., Midelfort, K. S., & Wittrup, K. D. (2000). Directed evolution of antibody fragments with monovalent femtomolar antigen-binding affinity. *Proceedings of the National Academy of Sciences of the United States of America, 97*, 10701–10705.

Bonazzi, M., Vasudevan, L., Mallet, A., Sachse, M., Sartori, A., Prevost, M. C., et al. (2011). Clathrin phosphorylation is required for actin recruitment at sites of bacterial adhesion and internalization. *The Journal of Cell Biology, 195*, 525–536.

Boulant, S., Kural, C., Zeeh, J. C., Ubelmann, F., & Kirchhausen, T. (2011). Actin dynamics counteract membrane tension during clathrin-mediated endocytosis. *Nature Cell Biology, 13*, 1124–1131.

Brenner, M. D., Zhou, R., & Ha, T. (2011). Forcing a connection: Impacts of single-molecule force spectroscopy on in vivo tension sensing. *Biopolymers, 95*, 332–344.

Brown, B., Li, C., Cheng, P., & Song, W. (1999). Trafficking of the Igα Igβ heterodimer with membrane Ig and bound antigen to the major histocompatibility complex class II peptide-loading compartment. *The Journal of Biological Chemistry, 274*, 11439–11446.

Busman-Sahay, K., Drake, L., Sitaram, A., Marks, M., & Drake, J. R. (2013). Cis and trans regulatory mechanisms control AP2-mediated B cell receptor endocytosis via select tyrosine-based motifs. *PLoS One, 8*, e54938.

Carrasco, Y. R., & Batista, F. D. (2006a). B-cell activation by membrane-bound antigens is facilitated by the interaction of VLA-4 with VCAM-1. *The EMBO Journal, 25*, 889–899.

Carrasco, Y., & Batista, F. D. (2006b). B cell recognition of membrane-bound antigen: An exquisite way of sensing ligands. *Current Opinion in Immunology*, *18*, 286–291.
Carrasco, Y. R., & Batista, F. D. (2007). B cells acquire particulate antigen in a macrophage-rich area at the boundary between the follicle and the subcapsular sinus of the lymph node. *Immunity*, *27*, 160–171.
Carrasco, Y. R., Fleire, S. J., Cameron, T., Dustin, M. L., & Batista, F. D. (2004). LFA-1/ICAM-1 interaction lowers the threshold of B cell activation by facilitating B cell adhesion and synapse formation. *Immunity*, *20*, 589–599.
Caswell, P. T., Vadrevu, S., & Norman, J. C. (2009). Integrins: Masters and slaves of endocytic transport. *Nature Reviews Molecular Cell Biology*, *10*, 843–853.
Chan, T. D., & Brink, R. (2012). Affinity-based selection and the germinal center response. *Immunological Reviews*, *247*, 11–23.
Chaturvedi, A., Dorward, D., & Pierce, S. K. (2008). The B cell receptor governs the subcellular location of Toll-like receptor 9 leading to hyperresponses to DNA-containing antigens. *Immunity*, *28*, 799–809.
Chaturvedi, A., Martz, R., Dorward, D., Waisberg, M., & Pierce, S. K. (2011). Endocytosed BCRs sequentially regulate MAPK and Akt signaling pathways from intracellular compartments. *Nature Immunology*, *12*, 1119–1126.
Chen, W., & Zhu, C. (2013). Mechanical regulation of T-cell functions. *Immunological Reviews*, *256*, 160–176.
Cocucci, E., Aguet, F., Boulant, S., & Kirchhausen, T. (2012). The first five seconds in the life of a clathrin-coated pit. *Cell*, *150*, 495–507.
Cureton, D. K., Massol, R. H., Whelan, S. P. J., & Kirchhausen, T. (2010). The length of vesicular stomatitis virus particles dictates a need for actin assembly during clathrin-dependent endocytosis. *PLoS Pathogens*, *6*, e1001127.
Cyster, J. G. (2010). B cell follicles and antigen encounters of the third kind. *Nature Immunology*, *11*, 989.
Dal Porto, J. M., Haberman, A. M., Kelsoe, G., & Shlomchik, M. J. (2002). Very low affinity B cells form germinal centers, become memory B cells, and participate in secondary immune responses when higher affinity competition is reduced. *The Journal of Experimental Medicine*, *195*, 1215–1221.
Davis, M. M., Krogsgaard, M., Huse, M., Huppa, J. B., Lillemeier, B. F., & Li, Q.-J. (2007). T cells as a self-referential, sensory organ. *Annual Review of Immunology*, *25*, 681–695.
Depoil, D., Fleire, S., Treanor, B. L., Weber, M., Harwood, N. E., Marchbank, K. L., et al. (2008). CD19 is essential for B cell activation by promoting B cell receptor-antigen microcluster formation in response to membrane-bound ligand. *Nature Immunology*, *9*, 63–72.
Depoil, D., Zaru, R., Guiraud, M., Chauveau, A., Harriague, J., Bismuth, G., et al. (2005). Immunological synapses are versatile structures enabling selective T cell polarization. *Immunity*, *22*, 185–194.
Derényi, I., Jülicher, F., & Prost, J. (2002). Formation and interaction of membrane tubes. *Physical Review Letters*, *88*, 238101.
Doherty, G. J., & Mcmahon, H. T. (2009). Mechanisms of endocytosis. *Annual Review of Biochemistry*, *78*, 857–902.
Dudko, O. K., Hummer, G., & Szabo, A. (2008). Theory, analysis, and interpretation of single-molecule force spectroscopy experiments. *Proceedings of the National Academy of Sciences*, *105*, 15755–15760.
Dustin, M. L., Bromley, S. K., Davis, M. M., & Zhu, C. (2001). Identification of self through two-dimensional chemistry and synapses. *Annual Review of Cell and Developmental Biology*, *17*, 133–157.
Dykstra, M. L., Longnecker, R., & Pierce, S. K. (2001). Epstein-Barr virus coopts lipid rafts to block the signaling and antigen transport functions of the BCR. *Immunity*, *14*, 57–67.

Evans, E., & Calderwood, D. A. (2007). Forces and bond dynamics in cell adhesion. *Science, 316*, 1148–1153.

Fleire, S. J., Goldman, J. P., Carrasco, Y. R., Weber, M., Bray, D., & Batista, F. D. (2006). B cell ligand discrimination through a spreading and contraction response. *Science, 312*, 738–741.

Freeman, S. A., Lei, V., Dang-Lawson, M., Mizuno, K., Roskelley, C. D., & Gold, M. R. (2011). Cofilin-mediated F-actin severing is regulated by the Rap GTPase and controls the cytoskeletal dynamics that drive lymphocyte spreading and BCR microcluster formation. *The Journal of Immunology, 187*, 5887–5900.

Friddle, R. W., Noy, A., & De Yoreo, J. J. (2012). Interpreting the widespread nonlinear force spectra of intermolecular bonds. *Proceedings of the National Academy of Sciences, 109*, 13573–13578.

Gauthier, N. C., Masters, T. A., & Sheetz, M. P. (2012). Mechanical feedback between membrane tension and dynamics. *Trends in Cell Biology, 22*, 527–535.

Gazumyan, A., Reichlin, A., & Nussenzweig, M. C. (2006). Ig beta tyrosine residues contribute to the control of B cell receptor signaling by regulating receptor internalization. *The Journal of Experimental Medicine, 203*, 1785–1794.

Giannone, G., Dubin-Thaler, B. J., Rossier, O., Cai, Y., Chaga, O., Jiang, G., et al. (2007). Lamellipodial actin mechanically links myosin activity with adhesion-site formation. *Cell, 128*, 561–575.

Gomez, G. A., McLachlan, R. W., & Yap, A. S. (2011). Productive tension: Force-sensing and homeostasis of cell-cell junctions. *Trends in Cell Biology, 21*, 499–505.

Gonzalez, S. F., Lukacs-Kornek, V., Kuligowski, M. P., Pitcher, L. A., Degn, S. E., Kim, Y.-A., et al. (2010). Capture of influenza by medullary dendritic cells via SIGN-R1 is essential for humoral immunity in draining lymph nodes. *Nature Immunology, 11*, 427–434.

Greenleaf, W. J., Woodside, M. T., & Block, S. M. (2007). High-Resolution, Single-Molecule Measurements of Biomolecular Motion. *Annual Review of Biophysics and Biomolecular Structure, 36*, 171–190.

Gretz, J. E., Norbury, C. C., Anderson, A. O., Proudfoot, A. E., & Shaw, S. (2000). Lymph-borne chemokines and other low molecular weight molecules reach high endothelial venules via specialized conduits while a functional barrier limits access to the lymphocyte microenvironments in lymph node cortex. *The Journal of Experimental Medicine, 192*, 1425–1440.

Griffiths, G. M., Tsun, A., & Stinchcombe, J. C. (2010). The immunological synapse: A focal point for endocytosis and exocytosis. *The Journal of Cell Biology, 189*, 399–406.

Harwood, N. E., & Batista, F. D. (2011). The cytoskeleton coordinates the early events of B-cell activation. *Cold Spring Harbor Perspectives in Biology. 3.* http://dx.doi.org/10.1101/cshperspect.a002360.

Heesters, B. A., Chatterjee, P., Kim, Y.-A., Gonzalez, S. F., Kuligowski, M. P., Kirchhausen, T., et al. (2013). Endocytosis and recycling of immune complexes by follicular dendritic cells enhances B cell antigen binding and activation. *Immunity.* http://dx.doi.org/10.1016/j.immuni.2013.02.023.

Hou, P., Araujo, E., Zhao, T., Zhang, M., Massenburg, D., Veselits, M., et al. (2006). B cell antigen receptor signaling and internalization are mutually exclusive events. *PLoS Biology, 4*, e200.

Hu, J., Lipowsky, R., & Weikl, T. R. (2013). Binding constants of membrane-anchored receptors and ligands depend strongly on the nanoscale roughness of membranes. *Proceedings of the National Academy of Sciences, 110*, 15283–15288.

Huang, J., Zarnitsyna, V. I., Liu, B., Edwards, L. J., Jiang, N., Evavold, B. D., et al. (2010). The kinetics of two-dimensional TCR and pMHC interactions determine T-cell responsiveness. *Nature, 464*, 932–936.

Huppa, J. B., Axmann, M., Mörtelmaier, M. A., Lillemeier, B. F., Newell, E. W., Brameshuber, M., et al. (2010). TCR-peptide-MHC interactions in situ show accelerated kinetics and increased affinity. *Nature, 463*, 963–967.

Husson, J., Chemin, K., Bohineust, A., Hivroz, C., & Henry, N. (2011). Force generation upon T cell receptor engagement. *PLoS One, 6*, e19680.

Jacob, M., Todd, L., Sampson, M. F., & Puré, E. (2008). Dual role of Cbl links critical events in BCR endocytosis. *International Immunology, 20*, 485–497.

Jang, C., Machtaler, S., & Matsuuchi, L. (2010). The role of Ig-α/β in B cell antigen receptor internalization. *Immunology Letters, 134*, 75–82.

Jimenez, R., Salazar, G., Yin, J., Joo, T., & Romesberg, F. E. (2004). Protein dynamics and the immunological evolution of molecular recognition. *Proceedings of the National Academy of Sciences of the United States of America, 101*, 3803–3808.

Junt, T., Moseman, E., Iannacone, M., Massberg, S., Lang, P., Boes, M., et al. (2007). Subcapsular sinus macrophages in lymph nodes clear lymph-borne viruses and present them to antiviral B cells. *Nature, 450*, 110.

Kaksonen, M., Toret, C. P., & Drubin, D. G. (2006). Harnessing actin dynamics for clathrin-mediated endocytosis. *Nature Reviews. Molecular Cell Biology*, 7, 404–414.

Katletz, S., Stroh, C., Rankl, C., Titulaer, U. M., & Hinterdorfer, P. (2010). Force-induced lysozyme—HyHEL5 antibody dissociation and its analysis by means of a cooperative binding model. *Biophysical Journal, 99*, 323–332.

Kim, S. T., Shin, Y., Brazin, K., Mallis, R. J., Sun, Z.-Y. J., Wagner, G., et al. (2012). TCR mechanobiology: Torques and tunable structures linked to early T cell signaling. *Frontiers in Immunology, 3*, 76.

Kim, J., Zhang, C.-Z., Zhang, X., & Springer, T. A. (2010). A mechanically stabilized receptor-ligand flex-bond important in the vasculature. *Nature, 466*, 992–995.

Klein, F., Diskin, R., Scheid, J. F., Gaebler, C., Mouquet, H., Georgiev, I. S., et al. (2013). Somatic mutations of the immunoglobulin framework are generally required for broad and potent HIV-1 neutralization. *Cell, 153*, 126–138.

Klotzsch, E., & Schutz, G. J. (2013). Improved ligand discrimination by force-induced unbinding of the T cell receptor from peptide-MHC. *Biophysical Journal, 104*, 1670–1675.

Kocks, C., & Rajewsky, K. (1988). Stepwise intraclonal maturation of antibody affinity through somatic hypermutation. *Proceedings of the National Academy of Sciences of the United States of America, 85*, 8206–8210.

Kouskoff, V., Famiglietti, S., Lacaud, G., Lang, P., Rider, J. E., Kay, B. K., et al. (1998). Antigens varying in affinity for the B cell receptor induce differential B lymphocyte responses. *The Journal of Experimental Medicine, 188*, 1453–1464.

Krieg, M., Helenius, J., Heisenberg, C.-P., & Müller, D. J. (2008). A bond for a lifetime: Employing membrane nanotubes from living cells to determine receptor-ligand kinetics. *Angewandte Chemie International Edition in English, 47*, 9775–9777.

Kuo, J. C., Han, X., Hsiao, C. T., Yates, J. R., & Waterman, C. M. (2011). Analysis of the myosin-II-responsive focal adhesion proteome reveals a role for β-Pix in negative regulation of focal adhesion maturation. *Nature Cell Biology, 13*, 383–393.

Lecuit, T., Lenne, P.-F., & Munro, E. (2011). Force generation, transmission, and integration during cell and tissue morphogenesis. *Annual Review of Cell and Developmental Biology, 27*, 157–184.

Le Trong, I., Aprikian, P., Kidd, B. A., Forero-Shelton, M., Tchesnokova, V., Rajagopal, P., et al. (2010). Structural basis for mechanical force regulation of the adhesin FimH via finger trap-like beta sheet twisting. *Cell, 141*, 645–655.

Levayer, R., Pelissier-Monier, A., & Lecuit, T. (2011). Spatial regulation of Dia and Myosin-II by RhoGEF2 controls initiation of E-cadherin endocytosis during epithelial morphogenesis. *Nature Cell Biology, 13*, 529–540.

Lingwood, D., McTamney, P. M., Yassine, H. M., Whittle, J. R. R., Guo, X., Boyington, J. C., et al. (2012). Structural and genetic basis for development of broadly neutralizing influenza antibodies. *Nature, 489*, 566–570.

Liu, W., Meckel, T., Tolar, P., Sohn, H. W., & Pierce, S. K. (2010). Antigen affinity discrimination is an intrinsic function of the B cell receptor. *The Journal of Experimental Medicine, 207*, 1095–1111.

Ma, Z., & Finkel, T. H. (2010). T cell receptor triggering by force. *Trends in Immunology, 31*, 1–6.

Ma, H., Yankee, T., Hu, J., Asai, D., Harrison, M., & Geahlen, R. (2001). Visualization of Syk-antigen receptor interactions using green fluorescent protein: Differential roles for Syk and Lyn in the regulation of receptor capping and internalization. *The Journal of Immunology, 166*, 1507–1516.

Malhotra, S., Kovats, S., Zhang, W., & Coggeshall, K. M. (2009). Vav and Rac activation in B cell antigen receptor endocytosis involves Vav recruitment to the adapter protein LAB. *Journal of Biological Chemistry, 284*, 36202–36212.

Manivel, V., Sahoo, N. C., Salunke, D. M., & Rao, K. V. S. (2000). Maturation of an antibody response is governed by modulations in flexibility of the antigen-combining site. *Immunity, 13*, 611–620.

Marshall, B. T., Long, M., Piper, J. W., Yago, T., McEver, R. P., & Zhu, C. (2003). Direct observation of catch bonds involving cell-adhesion molecules. *Nature, 423*, 190–193.

Mattila, P. K., Feest, C., Depoil, D., Treanor, B., Montaner, B., Otipoby, K. L., et al. (2013). The actin and tetraspanin networks organize receptor nanoclusters to regulate B cell receptor-mediated signaling. *Immunity*. http://dx.doi.org/10.1016/j.immuni.2012.11.019.

McKeithan, T. W. (1995). Kinetic proofreading in T-cell receptor signal transduction. *Proceedings of the National Academy of Sciences of the United States of America, 92*, 5042–5046.

Mcmahon, H. T., & Boucrot, E. (2011). Molecular mechanism and physiological functions of clathrin-mediated endocytosis. *Nature Reviews. Molecular Cell Biology, 12*, 517–533.

Moreno, N., Chevalier, M., Ronzon, F., Manin, C., Dupuy, M., Krell, T., et al. (2011). Unbinding forces of single pertussis toxin-antibody complexes measured by atomic force spectroscopy correlate with their dissociation rates determined by surface plasmon resonance. *Journal of Molecular Recognition, 24*, 1105–1114.

Morfill, J., Blank, K., Zahnd, C., Luginbühl, B., Kühner, F., Gottschalk, K.-E., et al. (2007). Affinity-matured recombinant antibody fragments analyzed by single-molecule force spectroscopy. *Biophysical Journal, 93*, 3583–3590.

Morfill, J., Neumann, J., Blank, K., Steinbach, U., Puchner, E. M., Gottschalk, K.-E., et al. (2008). Force-based analysis of multidimensional energy landscapes: Application of dynamic force spectroscopy and steered molecular dynamics simulations to an antibody fragment-peptide complex. *Journal of Molecular Biology, 381*, 1253–1266.

Natkanski, E., Lee, W.-Y., Mistry, B., Casal, A., Molloy, J. E., & Tolar, P. (2013). B cells use mechanical energy to discriminate antigen affinities. *Science, 340*, 1587–1590.

Nowak, S. A., & Chou, T. (2010). Models of dynamic extraction of lipid tethers from cell membranes. *Physical Biology, 7*, 026002.

Oda, M. (2004). Antibody flexibility observed in antigen binding and its subsequent signaling. *Journal of Biological Macromolecules, 4*, 45–56.

O'Neill, S. K., Veselits, M. L., Zhang, M., Labno, C., Cao, Y., Finnegan, A., et al. (2009). Endocytic sequestration of the B cell antigen receptor and toll-like receptor 9 in anergic cells. *Proceedings of the National Academy of Sciences, 106*, 6262–6267.

Pape, K. A., Catron, D. M., Itano, A. A., & Jenkins, M. K. (2007). The humoral immune response is initiated in lymph nodes by B cells that acquire soluble antigen directly in the follicles. *Immunity, 26*, 491–502.

Paus, D., Phan, T. G., Chan, T. D., Gardam, S., Basten, A., & Brink, R. (2006). Antigen recognition strength regulates the choice between extrafollicular plasma cell and germinal center B cell differentiation. *The Journal of Experimental Medicine, 203*, 1081–1091.

Phan, T. G., Grigorova, I., Okada, T., & Cyster, J. G. (2007). Subcapsular encounter and complement-dependent transport of immune complexes by lymph node B cells. *Nature Immunology*, *8*, 992–1000.

Phan, T. G., Paus, D., Chan, T. D., Turner, M. L., Nutt, S. L., Basten, A., et al. (2006). High affinity germinal center B cells are actively selected into the plasma cell compartment. *The Journal of Experimental Medicine*, *203*, 2419–2424.

Pierce, S. K., & Liu, W. (2010). The tipping points in the initiation of B cell signalling: How small changes make big differences. *Nature Reviews. Immunology*, *10*, 767–777.

Pierse, C. A., & Dudko, O. K. (2013). Kinetics and energetics of biomolecular folding and binding. *Biophysical Journal*, *105*, L19–L22.

Pitulescu, M. E., & Adams, R. H. (2010). Eph/ephrin molecules–a hub for signaling and endocytosis. *Genes & Development*, *24*, 2480–2492.

Qi, H., Egen, J. G., Huang, A. Y. C., & Germain, R. N. (2006). Extrafollicular activation of lymph node B cells by antigen-bearing dendritic cells. *Science*, *312*, 1672–1676.

Qureshi, O. S., Zheng, Y., Nakamura, K., Attridge, K., Manzotti, C., Schmidt, E. M., et al. (2011). Trans-endocytosis of CD80 and CD86: A molecular basis for the cell-extrinsic function of CTLA-4. *Science*, *332*, 600–603.

Rakshit, S., Zhang, Y., Manibog, K., Shafraz, O., & Sivasankar, S. (2012). Ideal, catch, and slip bonds in cadherin adhesion. *Proceedings of the National Academy of Sciences*, *109*, 18815–18820.

Ros, R., Schwesinger, F., Anselmetti, D., Kubon, M., Schäfer, R., Pluckthun, A., et al. (1998). Antigen binding forces of individually addressed single-chain Fv antibody molecules. *Proceedings of the National Academy of Sciences of the United States of America*, *95*, 7402–7405.

Saci, A., & Carpenter, C. L. (2005). RhoA GTPase Regulates B Cell Receptor Signaling. *Molecular Cell*, *17*, 205–214.

Sagawa, T., Oda, M., Ishimura, M., Furukawa, K., & Azuma, T. (2003). Thermodynamic and kinetic aspects of antibody evolution during the immune response to hapten. *Molecular Immunology*, *39*, 801–808.

Schmidt, A. G., Xu, H., Khan, A. R., O'Donnell, T., Khurana, S., King, L. R., et al. (2013). Preconfiguration of the antigen-binding site during affinity maturation of a broadly neutralizing influenza virus antibody. *Proceedings of the National Academy of Sciences*, *110*, 264–269.

Schnyder, T., Castello, A., Feest, C., Harwood, N. E., Oellerich, T., Urlaub, H., et al. (2011). B cell receptor-mediated antigen gathering requires ubiquitin ligase Cbl and adaptors Grb2 and Dok-3 to recruit dynein to the signaling microcluster. *Immunity*, *34*, 905–918.

Schwesinger, F., Ros, R., Strunz, T., Anselmetti, D., Güntherodt, H. J., Honegger, A., et al. (2000). Unbinding forces of single antibody-antigen complexes correlate with their thermal dissociation rates. *Proceedings of the National Academy of Sciences of the United States of America*, *97*, 9972–9977.

Schwickert, T. A., Lindquist, R. L., Shakhar, G., Livshits, G., Skokos, D., Kosco-Vilbois, M. H., et al. (2007). In vivo imaging of germinal centres reveals a dynamic open structure. *Nature*, *446*, 83–87.

Schwickert, T. A., Victora, G. D., Fooksman, D. R., Kamphorst, A. O., Mugnier, M. R., Gitlin, A. D., et al. (2011). A dynamic T cell-limited checkpoint regulates affinity-dependent B cell entry into the germinal center. *The Journal of Experimental Medicine*, *208*, 1243–1252.

Sharma, S., Orlowski, G., & Song, W. (2009). Btk regulates B cell receptor-mediated antigen processing and presentation by controlling actin cytoskeleton dynamics in B cells. *The Journal of Immunology*, *182*, 329–339.

Shih, T.-A. Y., Meffre, E., Roederer, M., & Nussenzweig, M. C. (2002). Role of BCR affinity in T cell dependent antibody responses in vivo. *Nature Immunology*, *3*, 570–575.
Shih, T.-A. Y., Roederer, M., & Nussenzweig, M. C. (2002). Role of antigen receptor affinity in T cell-independent antibody responses in vivo. *Nature Immunology*, *3*, 399–406.
Sixt, M., Kanazawa, N., Selg, M., Samson, T., Roos, G., Reinhardt, D. P., et al. (2005). The conduit system transports soluble antigens from the afferent lymph to resident dendritic cells in the T cell area of the lymph node. *Immunity*, *22*, 19–29.
Sohn, H. W., Gu, H., & Pierce, S. K. (2003). Cbl-b negatively regulates B cell antigen receptor signaling in mature B cells through ubiquitination of the tyrosine kinase Syk. *The Journal of Experimental Medicine*, *197*, 1511–1524.
Sohn, H. W., Tolar, P., & Pierce, S. K. (2008). Membrane heterogeneities in the formation of B cell receptor-Lyn kinase microclusters and the immune synapse. *The Journal of Cell Biology*, *182*, 367–379.
Stachowiak, J. C., Brodsky, F. M., & Miller, E. A. (2013). A cost–benefit analysis of the physical mechanisms of membrane curvature. *Nature Cell Biology*, *15*, 1019–1027.
Steidl, S., Ratsch, O., Brocks, B., Dürr, M., & Thomassen-Wolf, E. (2008). In vitro affinity maturation of human GM-CSF antibodies by targeted CDR-diversification. *Molecular Immunology*, *46*, 135–144.
Stoddart, A., Dykstra, M. L., Brown, B. K., Song, W., Pierce, S. K., & Brodsky, F. M. (2002). Lipid rafts unite signaling cascades with clathrin to regulate BCR internalization. *Immunity*, *17*, 451–462.
Stoddart, A., Jackson, A. P., & Brodsky, F. M. (2005). Plasticity of B cell receptor internalization upon conditional depletion of clathrin. *Molecular Biology of the Cell*, *16*, 2339–2348.
Sulchek, T., Friddle, R. W., & Noy, A. (2006). Strength of multiple parallel biological bonds. *Biophysical Journal*, *90*, 4686–4691.
Suzuki, Y., & Dudko, O. K. (2011). Biomolecules under mechanical stress: A simple mechanism of complex behavior. *Journal of Chemical Physics*, *134*, 065102.
Suzuki, K., Grigorova, I., Phan, T. G., Kelly, L. M., & Cyster, J. G. (2009). Visualizing B cell capture of cognate antigen from follicular dendritic cells. *The Journal of Experimental Medicine*, *206*, 1485–1493.
Tarlinton, D., & Good-Jacobson, K. (2013). Diversity among memory B cells: Origin, consequences, and utility. *Science*, *341*, 1205–1211.
Taylor, M. J., Perrais, D., & Merrifield, C. J. (2011). A high precision survey of the molecular dynamics of mammalian clathrin-mediated endocytosis. *PLoS Biology*, *9*, e1000604.
Thomas, W. (2008). Catch bonds in adhesion. *Annual Review of Biomedical Engineering*, *10*, 39–57.
Thorpe, I. F., & Brooks, C. L. (2007). Molecular evolution of affinity and flexibility in the immune system. *Proceedings of the National Academy of Sciences*, *104*, 8821–8826.
Tolar, P. (2011). Inside the microcluster: Antigen receptor signalling viewed with molecular imaging tools. *Immunology*, *133*, 271–277.
Tolar, P., Hanna, J., Krueger, P. D., & Pierce, S. K. (2009). The constant region of the membrane immunoglobulin mediates B cell-receptor clustering and signaling in response to membrane antigens. *Immunity*, *30*, 44–55.
Tolar, P., & Pierce, S. K. (2010). A conformation-induced oligomerization model for B cell receptor microclustering and signaling. *Current Topics in Microbiology and Immunology*, *340*, 155–169.
Tolar, P., Sohn, H. W., Liu, W., & Pierce, S. K. (2009). The molecular assembly and organization of signaling active B-cell receptor oligomers. *Immunological Reviews*, *232*, 34–41.

Treanor, B., Depoil, D., Bruckbauer, A., & Batista, F. D. (2011). Dynamic cortical actin remodeling by ERM proteins controls BCR microcluster organization and integrity. *The Journal of Experimental Medicine, 208*, 1055–1068.

Treanor, B., Depoil, D., Gonzalez-Granja, A., Barral, P., Weber, M., Dushek, O., et al. (2010). The membrane skeleton controls diffusion dynamics and signaling through the B cell receptor. *Immunity, 32*, 187–199.

Tsai, T., Choi, Y., Ma, W., Pomerening, J., Tang, C., & Ferrell, J. (2008). Robust, tunable biological oscillations from interlinked positive and negative feedback loops. *Science, 321*, 126.

Vascotto, F., Lankar, D., Faure-Andre, G., Vargas, P., Diaz, J., Le Roux, D., et al. (2007). The actin-based motor protein myosin II regulates MHC class II trafficking and BCR-driven antigen presentation. *The Journal of Cell Biology, 176*, 1007–1019.

Victora, G. D., & Nussenzweig, M. C. (2012). Germinal centers. *Annual Review of Immunology, 30*, 429–457.

Victora, G. D., Schwickert, T. A., Fooksman, D. R., Kamphorst, A. O., Meyer-Hermann, M., Dustin, M. L., et al. (2010). Germinal center dynamics revealed by multiphoton microscopy with a photoactivatable fluorescent reporter. *Cell, 143*, 592–605.

Wan, Z., Zhang, S., Fan, Y., Liu, K., Du, F., Davey, A. M., et al. (2013). B cell activation is regulated by the stiffness properties of the substrate presenting the antigens. *The Journal of Immunology, 190*, 4661–4675.

Weber, M., Treanor, B., Depoil, D., Shinohara, H., Harwood, N. E., Hikida, M., et al. (2008). Phospholipase C-gamma2 and Vav cooperate within signaling microclusters to propagate B cell spreading in response to membrane-bound antigen. *The Journal of Experimental Medicine, 205*, 853–868.

Wedemayer, G. J., Patten, P. A., Wang, L. H., Schultz, P. G., & Stevens, R. C. (1997). Structural insights into the evolution of an antibody combining site. *Science, 276*, 1665–1669.

Willis, J. R., Briney, B. S., DeLuca, S. L., Crowe, J. E., & Meiler, J. (2013). Human germline antibody gene segments encode polyspecific antibodies. *PLoS Computational Biology, 9*, e1003045.

Wu, Y., Vendome, J., Shapiro, L., Ben-Shaul, A., & Honig, B. (2011). Transforming binding affinities from three dimensions to two with application to cadherin clustering. *Nature, 475*, 510–513.

Yakovenko, O., Sharma, S., Forero, M., Tchesnokova, V., Aprikian, P., Kidd, B., et al. (2008). FimH forms catch bonds that are enhanced by mechanical force due to allosteric regulation. *Journal of Biological Chemistry, 283*, 11596–11605.

Yang, J., & Reth, M. (2010). Oligomeric organization of the B-cell antigen receptor on resting cells. *Nature, 467*, 465–469.

Yuseff, M.-I. I., Pierobon, P., Reversat, A., & Lennon-Duménil, A.-M. (2013). How B cells capture, process and present antigens: A crucial role for cell polarity. *Nature Reviews. Immunology, 13*, 475–486.

Yuseff, M.-I. I., Reversat, A., Lankar, D., Diaz, J., Fanget, I., Pierobon, P., et al. (2011). Polarized secretion of lysosomes at the B cell synapse couples antigen extraction to processing and presentation. *Immunity, 35*, 361–374.

Zhang, M., Veselits, M., O'Neill, S., Hou, P., Reddi, A. L., Berlin, I., et al. (2007). Ubiquitinylation of Ig beta dictates the endocytic fate of the B cell antigen receptor. *The Journal of Immunology, 179*, 4435–4443.

Zhang, S., Zhou, X., Lang, R. A., & Guo, F. (2012). RhoA of the Rho family small GTPases is essential for B lymphocyte development. *PLoS One, 7*, e33773.

Zimmermann, J., Oakman, E. L., Thorpe, I. F., Shi, X., Abbyad, P., Brooks, C. L., et al. (2006). Antibody evolution constrains conformational heterogeneity by tailoring protein dynamics. *Proceedings of the National Academy of Sciences of the United States of America, 103*, 13722–13727.

CHAPTER FOUR

The Role of BCR Isotype in B-Cell Development and Activation

Elena Surova[*,†], **Hassan Jumaa**[*,†,‡,1]

[*]Spemann Graduate School of Biology and Medicine (SGBM), Albert Ludwigs University Freiburg, Freiburg, Germany

[†]Department of Molecular immunology, Faculty of Biology, University of Freiburg and Max Planck Institute of Immunobiology and Epigenetics, Freiburg, Germany

[‡]Department of Immunology, Ulm University, Ulm, Germany

[1]Corresponding author: e-mail address: hassan.jumaa@uni-ulm.de

Contents

Advances in Immunology, Volume 123
ISSN 0065-2776
http://dx.doi.org/10.1016/B978-0-12-800266-7.00003-0

Abstract

The development and function of B lymphocytes critically depend on the non-germline B-cell antigen receptor (BCR). In addition to the diverse antigen-recognition regions, whose coding sequences are generated by the somatic DNA rearrangement, the variety of the constant domains of the Heavy Chain (HC) portion contributes to the multiplicity of the BCR types. The functions of particular classes of the HC, particularly in the context of the membrane BCR, are not completely understood. The expression of the various classes of the HC correlates with the distinct stages of B-cell development, types of B-cell subsets, and their effector functions. In this chapter, we summarize and discuss the accumulated knowledge on the role of the μ, δ, and γ HC isotypes of the conventional and precursor BCR in B-cell differentiation, selection, and engagement with (auto)antigens.

1. INTRODUCTION

1.1. B-cell reactivity to self is an essential aspect of B-cell differentiation and homeostasis

The B-cell antigen receptor (BCR) is a heterotetramer of two identical heavy chains (HCs) and two identical light chains (LCs). The protein complex is located at the plasma membrane together with an Ig-α/Ig-β heterodimer, which is essential for signal transduction (Fig. 4.1A).

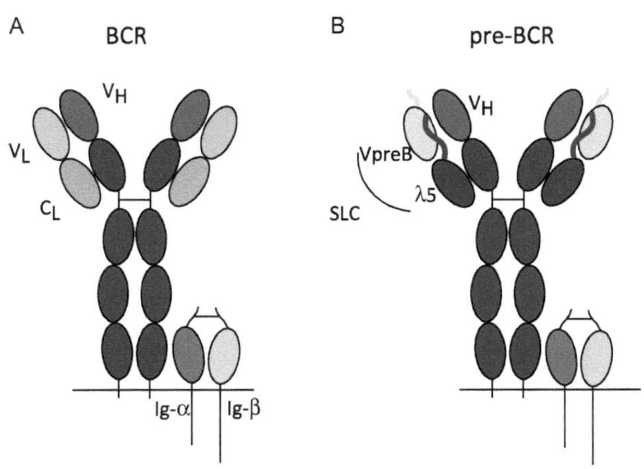

Figure 4.1 Schematic structure of a conventional IgM BCR (A) and a pre-BCR (B). Constant domains of the μ HC polypeptide are depicted in brown. (See the color plate.)

The diversity of the B-cell antigen receptors and their soluble forms, known as antibodies, is generated by a variety of mechanisms where the random recombination of the variable (V), diversity (D), and joining (J) gene segments plays a primary role. The VDJ joint in the heavy chain gene encodes the V_{HC} domain, while the VJ joint of the light chain gene encodes the V_{LC} domain.

The random nature of antigen receptor generation provides a powerful tool for recognizing an exceeding number of foreign antigens, yet the majority of the molecules available to the B cells originate from the organism's own tissues. A mechanism of immunoglobulin tolerance to the organism's self-tissues was discussed already in the works of Medawar and Burnet (Billingham, Brent, & Medawar, 1956; Burnet, 1959). Recent experimental studies show that mature B-cell repertoire in healthy individuals is not devoid of autoreactive antigen receptor specificities (Wardemann et al., 2003). Therefore, a further understanding of molecular events, regulating tolerance to self in the B-cell compartment, remains to be a critical focus of the current research. In the succeeding text, we will discuss the impact of self-recognition via antigen receptors expressed at the different stages of the B-cell development.

2. A PRECURSOR B-CELL RECEPTOR

The generation of the B-cell antigen receptor plays a central role in B-cell development. The V(D)J recombination of the heavy chain gene locus (IgH) precedes the recombination of the light chain genes. The productively rearranged IgH allele encodes a HC polypeptide, which assembles with the surrogate light-chain complex (SLC) as a precursor B-cell receptor (pre-BCR) (Melchers, 2005; Fig. 4.1B). The transport of the pre-BCR to the cell surface requires association with the Ig-α/Ig-β heterodimer (Papavasiliou, Misulovin, Suh, & Nussenzweig, 1995).

2.1. Structure and activation of the pre-B cell receptor

The overall organization of the pre-BCR Fab-like region resembles Fab of the conventional antigen receptor. Several elegant studies addressing the requirement of different pre-BCR components on pre-BCR assembly and pre-BCR function (Karasuyama, Kudo, & Melchers, 1990; Kitamura et al., 1992; Minegishi, Hendershot, & Conley, 1999; Ohnishi & Melchers, 2003; Tsubata & Reth, 1990) envisioned key aspects of the pre-BCR structure, which was recently resolved for the human pre-BCR.

The surrogate light chain is composed of two proteins, VpreB and λ5, that are homologous to the V_L and C_L portions of the conventional LC polypeptide, respectively. VpreB and λ5 form a tight complex, which does not involve covalent bonds (Bankovich et al., 2007; Melchers, 2005). VpreB is assembled in immunoglobulin (Ig) fold. While the V_L domain is composed of the seven antiparallel β-strands, VpreB provides only six β-strands, and the missing seventh β-strand is provided by the λ5 polypeptide (Minegishi et al., 1999).

In addition to the immunoglobulin (Ig) domains, both components of the SLC possess unique non-Ig regions (UR) that are enriched in charged amino acids and were proposed to participate in electrostatic intermolecular bonds (Ohnishi & Melchers, 2003). Positively charged arginines in the unique region of λ5 are evolutionary conserved and were shown to be essential for the proper pre-BCR function *in vitro* (Ohnishi & Melchers, 2003) and *in vivo* (Vettermann et al., 2008). In contrast to the LC–HC heterodimer, the SLC–HC complex has a loose structure (Bankovich et al., 2007). The complementarity determining region 3 (CDR3) of the SLC (CDR3-V_{SL}) differs substantially from the one of the conventional light chain as it is derived from the non-Ig parts of the SLC proteins. The N-terminus of the λ5 unique region is flexible and extends beyond the CDR3-H/SLC contact region. The unique region of VpreB is resolved and found to form a close contact with the variable domain of the heavy chain (Bankovich et al., 2007). The structural organization of the pre-BCR provided a precious insight in understanding the mode of its activity (see the succeeding text).

2.2. Importance of the precursor B-cell receptor

The expression of the pre-BCR marks the transition of pro-B cells to the immature B-cell stage. Available phenotypes of *in vivo* and *in vitro* pre-BCR-deficient models, as well as several clinical observations, gave rise to the current knowledge on the impact of the pre-B cell receptor on B-cell development.

The genetic deletion of the transmembrane and intracellular exons of the μHC, also known as μMT mutation, leads to an arrest of the B-cell development at the pro-/pre-B cell stage (Kitamura & Rajewsky, 1992). Ablation of surrogate light chain components, separately or in combination, leads to a severe block in B-cell development with a strong reduction in B-cell numbers in the periphery (Kitamura et al., 1992; Martensson, Argon, Melchers, Dul, & Martensson, 1999; Mundt, Licence, Shimizu, Melchers, & Martensson, 2001; Shimizu, Mundt, Licence, Melchers, & Martensson, 2002). A complete arrest of B-cell development was reported for the animals with a genetic deletion of several critical molecules involved in upstream signal transduction from the pre-BCR as Ig-α (Pelanda, Braun, Hobeika, Nussenzweig, & Reth, 2002), Syk (Cheng et al., 1995; Turner et al., 1995), or PI3K (Okkenhaug et al., 2002; Ramadani et al., 2010; Suzuki et al., 1999).

More detailed examination of the pre-BCR functions is based on *in vivo* and *in vitro* studies where the pre-BCR structure or signaling was partially perturbed. Mice lacking the λ5 component of the surrogate light chain have a severe reduction in the number of mature B cells. The development of the B cells in the bone marrow of λ5 knockout mice is characterized by the lack of the large pre-B cell population (Kitamura et al., 1992). Large pre-B cells undergo clonal proliferation (Karasuyama et al., 1994), and the lack of a large pre-B cell compartment reduces the output of B cells that continue the process of differentiation. Indeed, RAG-mediated assembly of the VDJ joint in the IgH gene locus in two out of three cases leads to the generation of frame-shifted transcripts, which do lead to the production of a functional HC polypeptide. Cell clones without a functional HC do not proceed in the development and perish.

Cell surface expression of the pre-BCR leads to the activation of several signaling cascades (Syk- and Akt-dependent) that modulate gene expression (downregulation of rag genes (Grawunder et al., 1995), dntt (Wasserman, Li, & Hardy, 1997), and surrogate light chain genes (Parker et al., 2005) and upregulation of SLP-65 (Schebesta, Pfeffer, & Busslinger, 2002)) and results in the termination of RAG-mediated IgH gene locus rearrangement. The accomplishment of the HC rearrangement program supports HC allelic exclusion and allows the progression of the cell to the rearrangement of the light chain gene loci after several rounds of cell division (Hess et al., 2001).

SLP-65 was identified as a critical adaptor molecule that supports the shift from proliferation to differentiation programs downstream of the pre-BCR (Herzog, Reth, & Jumaa, 2009; Jumaa et al., 1999). The loss of the SLP-65 leads to a compromised rearrangement of the LC genes (Herzog et al., 2008) and results in a sustained proliferation of pre-BCR-expressing cells and, ultimately, lymphomagenesis *in vivo* (Jumaa et al., 1999). A tumor suppression function of the SLP-65 regulates the differentiation program downstream of the pre-BCR. The SLP-65-dependent signaling complex leads to a stabilization of the FoxO transcription factors. This results in the reactivation of the rag genes and instigates the recombination of the LC gene loci (Herzog et al., 2008). The resulting LC polypeptide replaces the surrogate light chain from the complex with the HC, and thereby, the conventional B-cell antigen receptor is formed. Since a DNA template for the HC is generated by a random assembly of the

V (variable), D (diversity), and J (joining) gene segments, it was proposed that a pre-BCR serves an important quality control complex, where a productive HC can be tested for its compatibility to be assembled with a light chain (Keyna, Beck-Engeser, Jongstra, Applequist, & Jack, 1995).

2.3. The role of pre-BCR in V_H selection

Despite the random mechanism of VDJ joint formation, the HC repertoire at different stages of B-cell ontogeny and for the different B-cell subsets is characterized by biased V_H use (Perlmutter, Kearney, Chang, & Hood, 1985; Yancopoulos et al., 1984). Likewise, in fetal life and early bone marrow pro-B cells, the V_H repertoire is predominantly built up of V_H segments that are proximal to the D_H segments and belong to the V_H7183 and V_HQ52 families (Johnston, Wood, Bolland, & Corcoran, 2006). On the other hand, follicular B cells mainly use V_H segments that are distal to D_H, for example, V_HJ558 and V_H3609.

The likelihood for particular V_H segments to be found within the VDJ joint in the developing pro-B cell depends on the efficient recruitment of the recombination machinery. A number of genetic and epigenetic mechanisms contribute to the regulation of the accessibility of the V_H segments constituting the IgH gene locus (Choi et al., 2013; Subrahmanyam & Sen, 2012). Direct analysis of the HC transcripts and the ratio of in-frame to the out-of-frame VDJ rearrangements for different V_H gene families revealed that the shift from proximal to distal V_H use occurs at the transition from the pro-B to the late pre-B cell stages (Meng et al., 2011; ten Boekel, Melchers, & Rolink, 1997). Many of the heavy chains isolated before the pre-BCR expression stage were unable to pair with the SLC, while most sequences cloned from the late pre-B cells were able to form a pre-BCR. Therefore, biased V_H use beyond the pre-B cell receptor expression was essentially linked to the positive selection of the V_H by the pre-BCR (ten Boekel et al., 1997). Despite the variety of pre-BCR-deficient *in vivo* models, little information exists on the status of the V_H repertoire on a pre-BCR-deficient background. Contrary to the expectation, the use of V_H segments in the λ5-deficient mice was shown to have a normal bias towards distal V gene segments in splenic B cells (ten Boekel et al., 1997). Yet a novel mouse model with a block in B-cell development at the pre-B cell stage (a knock-in of N46Q mutation in the μHC gene) is the pre-BCR-deficient animal model, known to us, with a clear bias in the use of proximal over the distal V_H segments in splenic B cells (E. Surova and H. Jumaa, manuscript in preparation).

2.4. CDR3-H selection

In addition to the selection of the V_H sequences in the mature B-cell antigen receptor repertoire, the pre-BCR was proposed to be a quality control structure for the CDR3-H region (Martin et al., 2003). Indeed, there are approximately 100 functional V_H sequences annotated for the C57Bl/6j murine heavy-chain gene locus (Johnston et al., 2006); therefore, a quality control of the germ line-encoded V_H segments is a valid target for evolutionary selection. In contrast to the V_H, the CDR3-H sequences that are not encoded in the germline have no other possibility for selection but by the cell intrinsic mechanisms for quality control. The random nature of VDJ joint formation may potentially generate a heavy chain with a CDR3-H loop preventing proper folding of the HC and disturbance of the assembly with any light chain. Therefore, the ability to form a pre-BCR was proposed to be a checkpoint for the fitness of the HC being able to form part of the antigen receptor complex (Keyna, Beck-Engeser, Jongstra, Applequist, & Jack, 1995). Indeed, the pre-BCR structure revealed extensive contacts between the CDR3-H and a SLC (Bankovich et al., 2007), which provides support for the proposed hypothesis. Moreover, few heavy chains cloned from the bone marrow of λ5-deficient patients did not require LC/SLC binding for secretion when expressed in COS7 cells. Their CDR3-H sequences were enriched in positively charged amino acids (Minegishi & Conley, 2001). The higher use of positively charged amino acids in the CDR3-H was linked to the ability to recognize DNA (Radic & Weigert, 1994) and originally reported for the patients with systemic lupus erythematosus (SLE). In accord with data on human patients, SLC-null mice were reported to possess mature B cells with increased numbers of basic amino acid residues in the CDR3-H (Keenan et al., 2008). Moreover, SLC-deficient mice display an early appearance of ANA-reactive autoantibodies in the serum (Keenan et al., 2008). A pre-BCR-mediated influence on the structure of the CDR3-H was elegantly noted for the $V_H81 \times$ sequence in several studies. $V_H81 \times$ was found to be incapable of pairing with the SLC (Keyna, Applequist, Jongstra, Beck-Engeser, & Jack, 1995; ten Boekel et al., 1997). The few $V_H81 \times$ chains that could assemble with the SLC in a pre-BCR complex were found to have similar CDR3-H loop sizes, and 3D modeling revealed that these variable domains were superimposable in all regions, with CDR3-H taking a planar conformation (Martin et al., 2003). Pre-BCR-incompetent $V_H81 \times$ chains showed heterogeneity in the size and organization of the CDR3-H loop.

2.5. The role of the pre-BCR in HC allelic exclusion

Analyses of the early stages of B-cell development suggest that the precursor B-cell stage is important for establishing monoallelic HC expression. A well-recognized "feedback" model of HC allelic exclusion is supported by many experimental evidences (Vettermann & Schlissel, 2010). In particular, loss of the pre-BCR function in the μMT mutant mice leads to the high prevalence of early B cells expressing both HC alleles (Kitamura & Rajewsky, 1992). Intriguingly, genetic ablation of the SLC genes is dispensable for the HC allelic exclusion (Mundt et al., 2001; Shimizu et al., 2002). The seeming disagreement between the two studies was later resolved. In particular, it was shown that a strong selection for the SLC-independent HC expression on the SLC-null background results in the appearance of the μHCs that are capable to compensate the lack of the precursor B-cell receptor and establish HC allelic exclusion (Galler et al., 2004).

The molecular mechanisms ensuring the termination of the HC recombination process after the expression of the pre-BCR are described to depend on receptor signaling. In particular, the expression of the pre-BCR results in a downregulation of the genes associated with the recombination function (Grawunder et al., 1995; Wasserman et al., 1997) and leads to the progression of the cells into several rounds of proliferation, which results in ubiquitin-dependent degradation of RAG-2 protein (Hess et al., 2001; Jiang et al., 2005; Li, Dordai, Lee, & Desiderio, 1996; Lin & Desiderio, 1994).

2.6. Current models for pre-BCR engagement

Despite that the pre-BCR plays such a crucial role for the development of the B cells, biochemical and structural approaches to define the organization of the pre-BCR complex propose different mechanisms for its activation. In contrast to the case of the B-cell antigen receptor, which is activated after binding of its specific and unique antigen, several antigen-dependent (nonunique antigen) and cell-autonomous models were nominated to explain the molecular mechanism of pre-BCR engagement.

Galectin1 (Gal1), expressed by the bone marrow stromal cells, was proposed to function as a ligand for a human pre-BCR (Gauthier, Rossi, Roux, Termine, & Schiff, 2002), based on colocalization of signals for Gal1 and pre-BCR in an *in vitro* culture system. Furthermore, the interaction of the 24-amino acid-long fragment of the λ5 unique region with Gal1 was structurally resolved (Elantak et al., 2012). *In vivo* approaches to address the function of Gal1 as a pre-BCR ligand by analyzing B-cell development in Gal1 knockout mice did not confirm conclusions based on *in vitro* experiments (Espeli, Mancini, Breton, Poirier, & Schiff, 2009). No sign of compromised B-cell development was observed. In addition, a possible role of Gal1 as a murine pre-BCR did not receive any supportive evidence in another study (Bradl, Wittmann, Milius, Vettermann, & Jack, 2003). At the same time, heparan sulfate (HS) was proposed to serve as a murine pre-BCR ligand (Bradl et al., 2003), although experimental evidence was limited to the binding of recombinant pre-BCR molecules to the surface of the stromal cell line. In summary, the *in vitro* studies on Gal1 and HS as pre-BCR ligands lack convincing supportive *in vivo* evidence.

An alternative view of pre-BCR engagement proposed cell-autonomous pre-BCR activation (Rolink, Winkler, Melchers, & Andersson, 2000). The expression of the Fab-like fragment of the pre-BCR was demonstrated to form spontaneous dimers without any additional component (Bankovich et al., 2007), suggesting that pre-BCR intermolecular engagement might also occur on the cell surface. The advent of bone marrow-derived pro-B-cell lines allowed conditional reconstitution of pre-BCR components and downstream signaling molecules, which enabled the study of the pre-BCR activity in its native cell environment. Likewise, reconstitution of the pre-BCR components together with the conditionally inducible signal transduction molecule Slp65-ER^{T2} in Rag2$^{-/-}$λ5$^{-/-}$SLP65$^{-/-}$ (triple knockout) pro-B cells leads to signaling without the addition of a ligand (Meixlsperger et al., 2007).

All the different models of the mechanism of pre-BCR activation highlight the essential role of the λ5 non-Ig region, containing seven positively charged arginines, for proper pre-BCR function (Melchers, 2005). Flexibility of the λ5UR enriched in basic amino acid residues can result in general polyreactivity of the SLC (or the potential for a larger interactome) and an affinity for negatively charged protein surfaces. Indeed, in the absence of the rearranged HC gene, λ5 together with VpreB can be found at the cell surface, complexed with glycoproteins (Karasuyama, Rolink, & Melchers, 1993), where one of the partners was identified as BILL-cadherin (Ohnishi, Shimizu, Karasuyama, & Melchers, 2000). Binding of the λ5 to the glycan group attached to another protein, the μHC, was recently proposed to play an essential role in the pre-BCR function (Ubelhart et al., 2010). A cell-autonomous pre-BCR activation model proposed that negatively charged sialic acid-terminated glycan groups, attached to N46 of μHC C_H1, can mediate cross-linking of the neighboring pre-BCR complexes via λ5UR–glycan interaction *in trans*. Taking advantage of a powerful cell-based reconstitution system, assembly and autonomous ability to trigger

SLP-65-dependent intracellular Ca flux supported the requirement of the N46-linked glycan for proper pre-BCR formation and function *in vitro* (Ubelhart et al., 2010).

Both ligand-dependent and cell-autonomous models of pre-BCR engagement exclude the involvement of exogenous ligand/structure (non-self-ligand/structure) for the pre-BCR function. Therefore, the pre-B cell receptor is a bona fide autoreactive receptor. And early B lymphopoiesis is essentially linked to the reactivity to the self-derived structures.

2.7. Why is the SLC required for B-cell development?

Genetic deficiency of SLC components leads to the strong but not absolute block in B-cell development in mice, suggesting that some rescue mechanism can compensate for the lack of pre-BCR function. Developmental transition from the pre-B cell to the immature cell state manifests with the expression of conventional BCR, consisting of HC and LC polypeptides. Premature expression of an LC provides genetic rescue of B-cell development in λ5-deficient mice (Pelanda, Schaal, Torres, & Rajewsky, 1996) and rabbits with mutated IgH gene and deficiency in pre-BCR function (Robbins & Knight, 2011). Moreover, it was shown that LC gene loci in some cases can undergo rearrangement before the IgH gene (Novobrantseva et al., 1999). What kind of conventional BCRs can substitute the pre-BCR function?

A large fraction of newly generated immature lymphocytes bear self-reactive Ig receptors (Grandien, Fucs, Nobrega, Andersson, & Coutinho, 1994; Wardemann et al., 2003). The expression of an autoreactive BCR at the pre-B cell stage can mimic the signaling of the pre-BCR *in vitro* (Kohler et al., 2008). The expression of the autoreactive BCR in a pro-B cell line leads to the ligand-independent receptor signaling, recapitulating autonomous activity of the pre-BCR in the same type of experiment. Moreover, the non-Ig region of the λ5 polypeptide can be fully substituted by the CDR3-H loop of an autoreactive receptor. Functional redundancy between the pre-BCR and an autoreactive conventional B-cell receptor was demonstrated also *in vivo*. A block in B-cell development caused by the genetic deletion of the λ5 gene in mice can be rescued by the expression of the 3–83 BCR at pro-B-cell stage on H-2K^b autoreactive background (Eschbach et al., 2011).

There is a strong indication that the rescue of B-cell development on the SLC-null background in mice is driven by the autoreactive receptors. Mice with complete deficiency of SLC components have a strong block in B-cell development, but the residual lymphocytes that succeed in continuing differentiation bear autoreactive BCR specificities (Keenan et al., 2008). Prevalence of the autoreactivity may be explained by the selective advantage of the signals delivered by the autoreactive receptors. According to this view, on a wild-type background, the expression of the SLC brings an equal advantage to all SLC-compatible HCs by forming a self-reactive signaling receptor. In the absence of the SLC (and therefore pre-BCR), B cells bearing autoreactive immunoglobulin receptors are rescued by the signaling from their autoreactive BCR.

The lack of the pre-BCR function in the μMT mouse model, deficient for the membrane-bound form of the μHC, manifests with a strong block in B-cell differentiation (Kitamura & Rajewsky, 1992). Yet despite the lack of the functional pre-BCR, μMT mutation can support rudimentary B-cell development on C57Bl/6 (Macpherson et al., 2001) and even more proficiently on BALB/c backgrounds (Hasan, Polic, Bralic, Jonjic, & Rajewsky, 2002). The autoreactivity of the B cells, developing on the μMT background, was tested in mice with additional mutation in Fas (*lpr*). Intriguingly, animals with the μMT/*lpr* showed higher titers of chromatin autoantibodies than the *lpr* controls (Melamed, Miri, Leider, & Nemazee, 2000). Thus, both examples of the pre-BCR deficiencies, either from the lack of the SLC or due to the μMT mutation, demonstrate that in the absence of a pre-BCR-confined self-reactivity, the expression of the autoreactive BCR can support B-cell differentiation.

2.8. Are all HCs selected by the pre-BCR?

To what extent SLC- and LC-independent B-cell differentiation takes place on the wild-type background is not known. On the one hand, LC-independent expression of HC-containing, signaling-competent receptor complexes was reported in mice (Schuh, Meister, Roth, & Jack, 2003; Su et al., 2003) and human (Minegishi et al., 1998). On the other hand, LC-lacking HCs comprise a functional antibody subset in camelids (De Genst, Saerens, Muyldermans, & Conrath, 2006). Variable fragments of such heavy chains (named VHH) have structural constraints promoting better solubility and disfavoring binding of the light chains. The selection of these heavy chains is obscure but most likely is independent of the SLC. Interestingly, the expression of the chimeric transgenic constructs for the expression of llama VHH, fused to the constant part of the human μHC, can promote B-cell development in mice (Janssens et al., 2006). Such heavy chains coordinate the expression of the surface markers associated with progression through the pre-B cell stage, but do not induce rearrangement of the light-chain loci. In contrast to the light-chain binding Ig heavy chains, camelid VHH regions have characteristic hydrophobic-to-hydrophilic side chain substitutions in the framework regions, so that they are stable and soluble despite lacking the light chain. The structural determinants in mouse and human light-chain-independent heavy chains, which allow efficient folding, stability, and function, need to be identified.

An intriguing example of an unconventional heavy chain repertoire was recently reported for some of the bovine antibodies (Wang et al., 2013). Their heavy chains possess ultra-long CDR3-Hs regions, which assemble into two β-stranded stalk protrusions from the top of the V_{HC}–V_{LC} plane and terminate in an unusual disulfide-linked knob domain. According to the structural considerations of the human pre-BCR organization (Bankovich et al., 2007), it is likely that a bulky CDR3-H structure may not be compatible with the non-Ig region of the SLC. Therefore, a deeper analysis of the lymphocyte receptor repertoire across species calls for the new mechanisms of the Ig selection.

2.9. Are SLC functions exclusively associated with the pre-BCR?

If the function of the pre-BCR is redundant to the unavoidable naturally arising autoreactivity, it is conceivable that the strong B-cell deficiency on the SLC-null background has a complex nature, including disturbed function of the pre-B cell receptor and other complexes that belong to the SLC interaction network. The origin of this hypothesis dates back to the observation that VpreB and λ5 can be found on the surface of early B cells together with the μHC as well as in the absence of the μHC (Karasuyama et al., 1993, 1994; Misener, Downey, & Jongstra, 1991). In the absence of the μHC, the SLC components were shown to be a part of a glycoprotein complex, which also encompasses BILL-cadherin (Karasuyama et al., 1993; Ohnishi et al., 2000). *In vivo* genetic ablation of BILL-cadherin leads to a mild block in early B-cell development and a perturbed structure of the B-cell zone in the secondary lymphoid organs (Ohnishi, Melchers, & Shimizu, 2005). Yet it was not determined if the impact of BILL-cadherin on B-cell differentiation is B cell-intrinsic or B cell-extrinsic. A further indication for the potential role of the SLC aside from the pre-BCR complex originates from successful induction of Ca-flux in the μHC-negative pro-B-cell lines with λ5 cross-linking antibody (Misener et al., 1991). This suggests that pre-BCR-independent SLC-containing complex, found on the surface of developing B cells, may possess signaling properties.

A complex of calnexin with Ig-α and Ig-β signaling subunits of the pre-BCR was described to appear on the surface of the 38B9 transformed pro-B-cell line (Nagata et al., 1997). Moreover, it was reported that the application of the Ig-β -specific antibody to the RAG-deficient bone marrow cells can partially mimic the pre-BCR-mediated differentiation program (Nagata et al., 1997).

An expanded number of protein complexes, sharing structural and signaling components with the pre-B cell and B-cell antigen receptors, have been identified in the past decades. These studies were largely limited to the *in vitro* work, making it difficult to dissect the functional role of the discussed complexes in B-cell development.

3. SELF-RECOGNITION BY THE BCR

3.1. Receptor editing

The random pairing of the HC and the newly rearranged LC can lead to the assembly of an autoreactive BCR. The early attempts to probe the fate of self-reactive lymphocytes took advantage of the application of anti-Ig reactive sera to newborn animals in order to mimic autoantigen–BCR encounter (Cerny et al., 1986; Finkelman, Mond, & Metcalf, 1983; Gordon, Murgita, & Tomasi, 1975; Lawton, Asofsky, Hylton, & Cooper, 1972; Manning & Jutila, 1972; Takemori & Rajewsky, 1984; Tokuhisa, Gadus, & Herzenberg, 1981). Such treatments resulted in the specific depletion of B cells, depending on the specificity of the administered anti-Ig preparation. The advent of molecular techniques to generate genetically modified animals allowed to follow B-cell development upon the expression of defined self-specific BCRs. These pioneer studies of the BCR-transgenic mouse models unambiguously concluded that the expression of self-reactive antigen receptors has an adverse effect on B-cell development as it manifests with block in differentiation or loss of effector functions (Erikson et al., 1991; Goodnow et al., 1988; Nemazee & Buerki, 1989; Nemazee & Burki, 1989).

Paradoxically, the polyclonal repertoire of the newly generated immature B cells in mice and men comprises a high amount of autoreactive specificities (Grandien et al., 1994; Wardemann et al., 2003). Detailed analysis of peripheral B-cell populations revealed that innate-like B1 cells have high frequency of self-reactive BCRs (Pennell et al., 1989; Tornberg & Holmberg, 1995). Moreover, *in vivo* expression of the phosphatidylcholine-recognizing (α-PtC) and α-Thy1 transgenic BCRs supports the development of the innate type B1 and marginal zone B cells (Arnold, Pennell, McCray, & Clarke, 1994; Hayakawa et al., 1999; Wen et al., 2005).

Site-specific targeting of Ig genes encoding autoreactive BCRs within their native genetic loci revolutionized development of *in vivo* models to study B-cell autoreactivity. In particular, studies with knock-in design for the expression of autoreactive BCRs convincingly showed that editing of the self-reactive receptors can be a master mechanism for establishing the polyclonal B-cell repertoire and regulate the expression of self-reactive receptors (Casellas et al., 2001; Novobrantseva et al., 2005; Pelanda et al., 1997). Moreover, a single-cell receptor analysis of human B cells revealed that up to 75% of early immature B cells are autoreactive (Wardemann et al., 2003). A gradient decline in the fraction of the autoreactive B cells along the differentiation axis is in agreement with the contribution of the receptor editing in shaping the mature naive B-cell repertoire. Self-antigen-mediated revision of a BCR takes place in many of the generated Ig knock-in mouse models (Casellas et al., 2001; Doyle-Cooper et al., 2013; Hippen et al., 2005; Pelanda et al., 1997; Pewzner-Jung et al., 1998).

Revision of the autoreactive receptors at the immature B-cell stage is achieved via secondary rearrangement of the light chain gene loci (Nemazee & Weigert, 2000; Pelanda et al., 1997). This process is functionally redundant to the light chain gene rearrangement, initiated after expression of the pre-BCR. Signaling from the autoreactive BCRs could mimic signaling from the pre-BCR *in vitro* and could functionally replace it *in vivo* (Eschbach et al., 2011; Kohler et al., 2008). Thus, it is not surprising that autoreactivity at immature B-cell stage is found in healthy individuals. Furthermore, autoreactivity of immature cells can contribute to the establishment of variable BCR repertoire.

3.2. Tonic BCR signaling and encounter of autoantigens

Detailed molecular events of BCR engagement by autoantigens in a healthy individual remain unknown. The lack of an anatomical and functional compartment for BCR selection as a thymus for the TCR-expressing lymphocytes, leaves the hypothesis on organized B-cell selection unsupported. The possibility of Ig editing, taking place in B cells, but not in T cells (Kreslavsky et al., 2013), is proposed to be the major mechanism of the BCR diversification for the establishment of the polyclonal B-cell repertoire (Casellas et al., 2001; Novobrantseva et al., 2005). Moreover, a recent study suggests that encounter of autoantigens plays an important role in supporting mature stages of the B-cell differentiation and responsiveness (Zikherman, Parameswaran, & Weiss, 2012).

Tonic BCR signaling and encounter of low-affinity autoantigens by the BCR have not been clearly separated. BCR activity in the absence of obvious engagement by the foreign antigens was recognized important for the maintenance of mature B cells. Loss of surface BCR expression leads to major changes in the B-cell transcriptome and a cell withdrawal (Kraus, Alimzhanov, Rajewsky, & Rajewsky, 2004; Lam, Kuhn, & Rajewsky, 1997; Tze et al., 2005). Along with the BCR, BAFF-BAFF-R plays a critical role in survival of B cells in the periphery. Recent data suggest that BAFF-R activation is dependent on BCR surface expression and downstream signaling molecules, which reiterates the impact of BCR expression for the B-cell survival (Schweighoffer et al., 2013). While a conditional activation of a Syk kinase leads to terminal differentiation of the B cells (Hug, Hobeika, Reth, & Jumaa, 2013), mimicking a mild BCR signaling through the expression of viral protein LMP2A is able to support the development of the major mature B-cell subsets in the absence of the native BCR (Casola et al., 2004). Overall, the compelling amount of experimental evidence suggests that the presence of a BCR on resting B cells is critical for B-cell differentiation and maintenance.

3.3. Current models for BCR engagement

The functional biology of a BCR and its secreted form (antibody) is mainly studied in the context of their interaction with specific high-affinity antigen. Binding of the antigen by the BCR results in the activation of the multiple signaling cascades inside a B cell, but a definite understanding of the translation of the extracellular events to the intracellular compartment is lacking (Kurosaki, 2011). Several models aim to describe the earliest events of the BCR activation, and they are reviewed elsewhere (Treanor, 2012; Yang & Reth, 2010).

The major transition from resting to activation state of the BCR is thought to happen via a conformational change. Analysis of crystal structures of immunoglobulins in the absence and in complex with the antigens supports that resting and bound BCRs are present in different conformations. The flexibility of immunoglobulin structure was analyzed in a number of studies. Likewise, a conformational diversity is demonstrated for many unmutated antibodies upon interaction with the ligands, while affinity-maturated immunoglobulins display restricted conformational variability (Manivel, Sahoo, Salunke, & Rao, 2000; Wedemayer, Patten, Wang, Schultz, & Stevens, 1997). An example of the polyreactive SPE7 antibody suggests that polyreactivity is linked to a number of diverse conformations that were identified for this particular antibody (James, Roversi, & Tawfik, 2003). Molecular domains, responsible for the flexibility of binding various antigens, spread outside the hypervariable loops of the V_{HC} and V_{LC}. An example of the IgM rheumatoid factor (RF) in the complex with the Fc demonstrated that the actual antigen-binding site can extend far beyond the CDR region (Corper et al., 1997). This led to a suggestion that the increased accessibility of the "classical" site, formed by the CDR loops of the IgM RF antibody, might make the RF available for interaction with another antigen even after RF binding to the Fc.

The conformation change, associated with the binding of an antigen, affects not only the structure of antigen-binding variable fragment (Fab) of the immunoglobulin. A systematic analysis of conformational differences between the antigen-bound (Ag-bound) and Ag-free resolved immunoglobulin structures revealed that conformational change spreads beyond the variable part of the immunoglobulin (Sela-Culang, Alon, & Ofran, 2012). Likewise, apart from the flexible CDR3-H region, a conformational change affected the C_H1 domain and the orientation of the V-C domains of the Fab fragment. Moreover, structural determinants that are distant from the Ag-binding site influence binding of the Ag itself. The structure of the constant part of the immunoglobulin HC was shown to have a strong impact on the antigen binding (Cooper, Robertson, Granzow, & Greenspan, 1994; Morahan, Berek, & Miller, 1983; Morelock et al., 1994; Pritsch et al., 1996, 2000; Torres, Fernandez-Fuentes, Fiser, & Casadevall, 2007). Thus, the change of immunoglobulin affinity to the same antigen in the context of different isotypes is proposed to play an important role in the antigen-specific immune response. An isotype-restriction phenomena, describing a preferential generation of one antibody isotype upon challenge with a particular class of antigens, is proposed to be linked to a certain conformational advantage imposed by the HC/LC constant region on the ability of the Fab to bind the antigen (Greenspan & Cooper, 1993; Tongren, Corran, Jarra, Langhorne, & Riley, 2005).

Mutual impact on the function of the variable and constant parts of the soluble immunoglobulins is bringing important focus on the role of the immunoglobulin HC isotype for the antibody-mediated therapies (Liu et al., 2003; Tudor et al., 2012).

Distinction between the conformation of the resting and Ag-bound immunoglobulin suggests that cell-autonomous activity of the pre-BCR can be driven by the specific conformation of this receptor. The presence of the non-Ig region of λ5 is essential for mediating pre-BCR autonomous signaling (Ohnishi & Melchers, 2003). Yet it was not possible to define the structure of this part of the SLC (Bankovich et al., 2007). Therefore, it is conceivable that instable conformation of the SLC is responsible for the cell-autonomous pre-BCR activity.

Productively recombined IgH gene locus is expressed as a μHC in the context of the pre-BCR and BCR. B-cell maturation and terminal differentiation are manifested by the expression of other BCR isotypes. Therefore, "what is the impact of the heavy chain isotype as a part of the B-cell antigen receptor on later stages of B-cell development?" is an important question.

4. EXPRESSION OF IgD BCR ON MATURE B CELLS

4.1. Differences in structure suggest different functions for BCR isotypes

In addition to the immense variety of the antigen-binding fragments generated as a result of the somatic V(D)J recombination, the BCR is diversified by the use of the several classes of heavy chains. The expression of the γHC, εHC, and αHC classes occurs on antigen-experienced cells in the secondary lymphoid organs and requires class-switch recombination in the IgH gene locus. The predominant expression of the μHC and δHC isotypes marks developing and mature naive follicular B cells respectively.

The distribution of functions between the μHC and δHC isotypes in orchestrating B-cell differentiation is not fully understood. While μHC can be found both membrane-associated (IgM BCR) and soluble (IgM antibodies), δHC is predominantly a part of the surface receptor—IgD BCR. IgD antibodies are barely detectable in the serum of healthy primates and rodents. Therefore, putative effector functions of soluble IgD are not evident.

Pre-B cell and immature stages of B-cell development manifest with the expression of the μHC. IgM knockout animals show mild block in pre-B cell development (Lutz et al., 1998). *In vitro* assembly and function of the pre-BCR were shown to require μHC but not δHC (Ubelhart et al.,

2010). Upon rearrangement of the conventional light chain and expression of the non-autoreactive BCR, immature B cells migrate to the secondary lymphoid organs and begin to express IgD. Alternative splicing and class-switch recombination regulate the expression of the δHC in mammals.

The prevalent expression of the μHC over δHC at the early stages of B-cell development is reverted for mature B cells, where IgD becomes the dominant antigen receptor. An alternative mode of heavy chain isotype expression is found on mature innate-like B-cell subsets—B1 cells in the peritoneal and pleural cavities and marginal zone B cells in the spleen. These distinct populations predominantly express IgM on the surface and as secreted antigen receptor.

4.2. Structural and signaling differences between IgM and IgD

In contrast to the μHC, the δHC gene exhibits a high degree of structural plasticity and low conservation among vertebrate taxa (Edholm, Bengten, & Wilson, 2011; Rogers, Richardson, Scinicariello, & Attanasio, 2006; Fig. 4.2). The quantity of immunoglobulin domains among IgDs varies from two in mice to fourteen in fugu (Flajnik, 2002; Magadan-Mompo, Sanchez-Espinel, & Gambon-Deza, 2011).

The C_H1 and C_H2 domains of IgD are connected via a flexible hinge region, while C_H1 and C_H2 of the IgM lack a hinge. Fishes, on the other hand, lack the hinge region in Cδ.

Protein folding and protein–protein interaction are linked to protein glycosylation (Arnold, Wormald, Sim, Rudd, & Dwek, 2007; Dalziel, Crispin, Scanlan, Zitzmann, & Dwek, 2014). N- and O-linked glycosylation patterns of the δHC molecules strongly vary between the different phylogenetic groups (Rogers et al., 2006).

Signaling disparity between the IgM and IgD has been addressed in some experimental settings, but it still lacks the detailed analysis. The intracellular tail of the membrane-tethered forms of IgM and IgD is identical in mice and men and comprises only three amino acids Lys-Val-Lys (KVK). Therefore, both μHC and δHC isotypes require association with Ig-α/Ig-β heterodimer for transmitting the activation signals. Yet reconstitution of the NIP-specific receptor as IgM and IgD in cell lines revealed that the kinetics of phosphorylation of intracellular proteins and the mode of the antigen receptor activation differ between the IgM and IgD (Kim & Reth, 1995). *In vivo* stimulation with the antigen was shown to induce downmodulation of surface IgM receptors but not of IgD (Brink et al.,

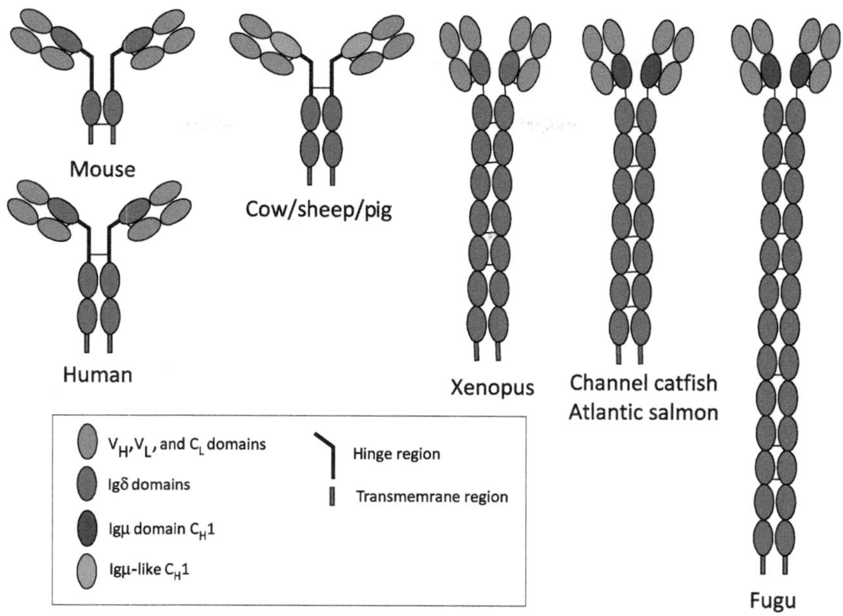

Figure 4.2 Plasticity of the structure of IgD across different species. The following resources were referred to infer the schematic structure of IgD BCR: human IgD (UniProtKB entry PO1880), mouse IgD (UniProtKB entry P01882), cow/sheep/pig IgD (Rogers et al., 2006; Zhao et al., 2002), channel catfish IgD (Wilson et al., 1997), *Xenopus* IgD (Ohta & Flajnik, 2006), Atlantic cod IgD (Stenvik & Jorgensen, 2000), and fugu IgD (Saha, Suetake, Kikuchi, & Suzuki, 2004). (See the color plate.)

1992; Carsetti, Kohler, & Lamers, 1993). Encounter of a cognate antigen by $IgM^{+}IgD^{neg}$ B cells leads to B-cell apoptosis, while $IgM^{+}IgD^{+}$ B cells are rescued from clonal deletion (Carsetti et al., 1993). According to *in vivo* and *in vitro* studies, signaling from IgD is important for efficient AID function (Roes & Rajewsky, 1993). The activation of IgM, on the other hand, was shown to downregulate AID levels (Hauser et al., 2008).

A detailed understanding, which molecular events discriminate IgM- and IgD-induced signaling, remains unknown. The Ig-α signaling subunit of the BCR is glycosylated differently when associated with either IgM or IgD (Pogue & Goodnow, 1994), although the functional properties of different Ig-α-glycosylated forms remain unknown.

4.3. Can IgD replace IgM in B-cell development?

Strict temporal hierarchy of the expression of the μ and δ heavy chains suggests nonoverlapping isotype-specific functions, although *in vivo* genetic models provided controversial evidences. Germline ablation of the membrane portion of μHC in mice (Kitamura & Rajewsky, 1992) and targeted deletion of part of the Cμ gene in rats (Menoret et al., 2010) result in strong block in B-cell development, and in both cases, the expression of the δHC is not supported. Severe agammaglobulinemia and block in B-cell development were reported for patients with mutations in μHC coding sequence (Yel et al., 1996). Upon complete deletion of the constant part of the μHC gene (Cμ), the coding region of the constant part of δHC (Cδ) reaches close proximity to the IgH intronic enhancer and is successfully expressed. In this animal model, δHC could rescue B-cell development in the absence of μHC (Lutz et al., 1998). Thus, surface expression of IgD instead of IgM can fully support differentiation of B2 cells *in vivo* but only partially promotes B1-cell development.

As IgD is the dominant antigen receptor on the surface of mature B2 cells, the lack of IgD has an adverse effect on the mature B2-cell compartment (Nitschke, Kosco, Kohler, & Lamers, 1993; Roes & Rajewsky, 1993). The expression of surface IgD is required for the optimal serological immune response. IgD knockout mice have compromised affinity maturation in the course of T-cell dependent immune response (Roes & Rajewsky, 1993).

Selective expression of either μHC or δHC was studied in transgenic animal models and allowed to compare HC isotypes in the context of the same V_H. It was shown that exclusive expression of the HEL IgD does not support the development of B1 and B2 cells as efficient as HEL IgM BCR, yet the most efficient B-cell development was observed in the HEL IgM–IgD transgenic mice (Brink et al., 1992). Upon stimulation with its cognate antigen, HEL IgM transgenic B cells are more sensitive than their HEL IgD counterparts *in vivo* and *in vitro* (Brink, Goodnow, & Basten, 1995; Brink et al., 1992). These findings support the hypothesis of nonoverlapping functions of IgM and IgD in sustaining B-cell development and function.

Preclusion of the highly self-reactive BCRs from the mature B-cell repertoire is achieved through the following mechanisms of receptor/cell selection. Clonal deletion and receptor editing are recognized as the major mechanisms of B-cell tolerance at the immature B-cell stage in the bone marrow (Pelanda & Torres, 2012), while anergy and deletion impose peripheral B-cell tolerance (Cambier, Gauld, Merrell, & Vilen, 2007). The differential involvement of the IgM and IgD in B-cell tolerance is still controversially discussed. In the following chapters, we summarize the differences between IgM and IgD BCR isotypes and highlight the potential impact and consequences of such differences.

4.4. Differences in signal initiation between IgM and IgD

The differential response of the IgM and IgD B-cell receptors was clearly noticed in the IgHEL transgenic mouse models. Continuous exposure to the soluble neo-self-antigen HEL prevents differentiation of transitional IgHEL transgenic B cells into follicular B-cell pool and manifests with downregulation of surface IgM but does not affect the expression of the IgD (Brink et al., 1995; Goodnow et al., 1988). Resistance to activation demonstrated by the peripheral IgHEL-expressing lymphocytes was the pioneering description of B-cell anergy. Interestingly, the same pattern of selective downmodulation of the IgM characterizes anergic B cells in humans (Quach et al., 2011). Moreover, analysis of human B cells, class-switched to IgD, identifies a high proportion of autoreactive receptors among them (Koelsch et al., 2007; Zheng et al., 2004). Thus, withdrawal of IgM from the cell surface of the B cells, while retaining IgD, is likely to identify B cells activated by the self-antigens. In order to explain resistance of the IgD to the antigen-mediated receptor internalization, a model of higher activation threshold was proposed (Brink et al., 1992).

4.5. IgD may fine-tune BCR responses to self-ligands in the periphery

Editing of the autoreactive specificity is recognized as the major mechanism of B-cell tolerance to membrane-bound autoantigens (Ait-Azzouzene et al., 2005; Gay, Saunders, Camper, & Weigert, 1993; Hippen et al., 2005; Lang et al., 1996; Pelanda et al., 1997; Tiegs, Russell, & Nemazee, 1993). Immature cells can alter their autoreactive BCR idiotype by replacing existing light or heavy chains (Chen, Nagy, Prak, & Weigert, 1995; Nemazee & Weigert, 2000; Pelanda et al., 1997; Tiegs et al., 1993). Signaling from the autoreactive BCRs is shown to mimic pre-BCR signaling (Kohler et al., 2008), and both types of receptors are assembled with the use of μHC isotype. High sensitivity of the immature cells to elimination of even ultra-low-affinity autoreactive receptors was appreciated in experiments with the 3–83 BCR-transgenic mice (Lang et al., 1996). The expression of the 3–83 BCR, selected in the germinal center of the H-2K^{d+} B cells from the BALB/c background as IgG1 (Ozato, Mayer, & Sachs, 1980), is efficiently edited at the immature B-cell stage even if the 3–83 BCR was expressed in the theoretically non-autoreactive, BALB/c background (Pelanda et al., 1997). This unexpected observation might be explained in multiple ways that include a decreased expression level of the 3–83 knock-in BCR, low interaction with the MHCI haplotype of the BALB/c background, or low-affinity interaction with unknown antigens that are available in the environment. The latter scenario suggests that an exceptionally high sensitivity of IgM BCR provides sufficient triggering for the induction of receptor editing, which removes a wide range of low-affinity self-reactive specificities from the immature B cells and at the same time diversifies the BCR repertoire. When tested in the same knock-in experimental approach but in IgM-deficient mice, the expression of the 3–83 BCR as an IgD isotype does not lead to receptor editing on H-2K^{d+} low autoantigenic background (E. Surova and H. Jumaa, manuscript in preparation). These results are in full agreement with the higher activation threshold for IgD BCR as compared with IgM BCR.

Taken together, it seems that while high sensitivity of the IgM BCR in early immature B-cell stage is beneficial, as it leads to the LC editing and receptor diversification, the expression of IgD with higher activation threshold may fine-tune B-cell responses to a variety of self-antigens available in the periphery (Fig. 4.3).

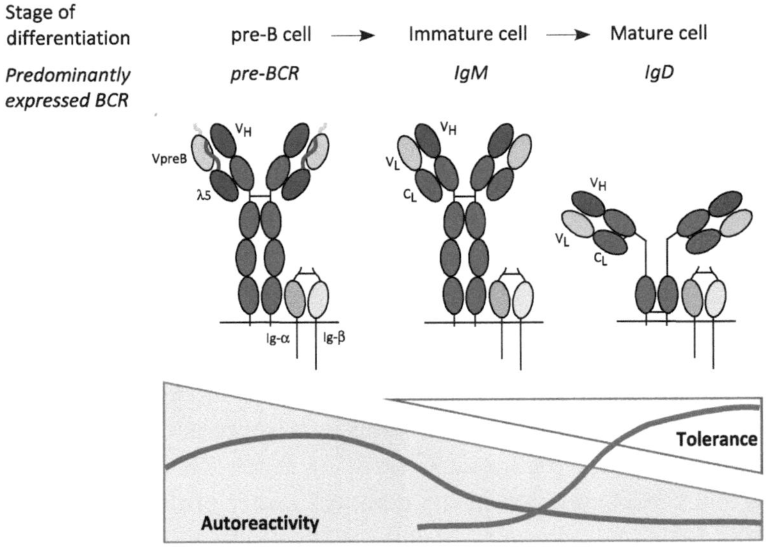

Figure 4.3 (*Top*) Schematic structure of murine pre-BCR and IgM and IgD receptors and the stages at which they are expressed. (*Bottom*) Occurrence of autoreactive (gray polygon) and tolerant (white triangle) BCRs during B-cell ontogeny from pre-B to mature B-cell stage. The brown line depicts the expression of μHC; the green line, δHC. (See the color plate.)

4.6. Potential determinants of IgM and IgD activation disparity

The reasons for a higher activation threshold for the IgD BCR are not completely understood. According to earlier studies on the assembly of the BCRs with the signal-transducing subunits, IgD could also be presented on the cell surface without Ig-α/Ig-β heterodimer via glycosylphosphatidylinositol linkage (Wienands & Reth, 1992). Thereby, such a complex of the IgD with its antigen should be sequestered from the signaling molecules. Moreover, the transmembrane form of the IgD antigen receptor imposes a distinct type of glycosylation on the Ig-α signaling subunit. In complex with IgD, Ig-α possesses an EndoH-resistant N-linked glycosylation, while in the complex with IgM, BCR Ig-α is modified by an EndoH-sensitive glycosylation (Pogue & Goodnow, 1994a,1994b). The functional impact of the different forms of Ig-α glycosylation remains unknown.

An organization of the receptors and associated complexes from both sides of the cell membrane attracts currently particular interest. To this date, the membrane-proximal intracellular environment of B-cell antigen receptors and its dynamics are not fully understood. The lipid microdomains surrounding the receptors and lipid-linked signaling play an important role in the BCR-mediated cell activation (Simons & Toomre, 2000; Werner, Hobeika, & Jumaa, 2010). An impact of membrane microdomains and organization of the molecular network proximal to the BCR for the different BCR isotypes remain to be determined.

An active involvement of intracellular actin cytoskeleton emerges as a critical regulator of BCR organization and signaling (Harwood & Batista, 2010). Analysis of the IgM and IgD BCRs on the plasma membrane of the resting B cells revealed that receptor complexes are organized into discrete clusters (Mattila et al., 2013). The stochastic mobility of the BCR measured by the diffusion coefficient in the resting lymphocytes is higher for the IgM receptor type than for IgD (Treanor et al., 2010). These pioneering results highlight the differences in the organization of IgM and IgD on naive B cells and support the previously accumulated knowledge on discrepancies between the functions of the IgM and IgD.

5. CLASS-SWITCHED BCRs

5.1. Zooming into IgG receptor isotypes

The expression of class-switched BCR mostly involves modification of the variable antigen-binding sequence, and it deprives the cell of expression of

IgM and IgD receptors. Therefore, it is important to understand how the modified BCR structure is adapted to support the maintenance of memory B cells in the absence and reactivation upon encounter of its cognate antigen.

The IgH gene locus downstream of the Cμ and Cδ coding regions in mammals contains also coding sequences for the γ, ε, and α classes of HCs. The diversity of the HC structures is implicated in conferring distinct functions to the secreted antibodies and membrane-bound Ig receptors. The diversified structural determinants of the various classes of HCs in the context of the surface BCR are found across both sides of the plasma membrane.

The murine membrane-bound forms of the γHCs and εHCs possess a conserved 28-amino acid intracellular tails, and αHC, 14 amino acids (Reth, 1992). The unique intracellular sequence is extended from the 3 amino acids present in IgM and IgD was proposed to have distinct signaling properties. A principal requirement of the intracellular tail of the IgG1 and IgE for mounting a robust antigen-mediated immune responses was demonstrated upon targeted shortening of intracellular part of IgG1 (Kaisho, Schwenk, & Rajewsky, 1997) and IgE (Achatz, Nitschke, & Lamers, 1997) intracellular tails. Apart from the impact of the intracellular tail of the BCR-mediated response, it was shown that the full-length intracellular tail of the IgG1 plays an important role in supporting surface BCR expression (Kaisho et al., 1997).

The structure of the extracellular portion of the IgG molecules is conserved between the IgG subclasses and across the species. The strongest conservation of all the parts of IgG species is attributed to the transmembrane and intracellular tails. The highest variability, on the other hand, characterizes the hinge region. For example, the hinge region of the human IgG contains different numbers of the disulfide interchain bonds: from two for IgG1 and IgG4 to four for IgG2 and up to eleven for IgG3 (Liu & May, 2012). Apart from the classical S–S bonds, it was recognized that IgG2 and particularly IgG4 can persist in forms with nonclassical disulfide bond structures. Intra- and interchain disulfide bonds play a critical role in immunoglobulin folding and stability of the antibodies. Likewise, it is conceivable that disulfide "code" affects not only the structure, stability, and effector functions of the soluble immunoglobulins but also a surface BCR.

5.2. The impact of IgG on B-cell responses

Stimulation of B cells expressing IgM or IgG types of BCR is characterized by the superior amount of the antibody production upon engagement of the

IgG (Horikawa et al., 2007; Martin & Goodnow, 2002). The intracellular part of the IgG plays a critical role in supporting stronger signaling (Engels et al., 2009) and plasma-cell differentiation in response to antigen encounter (Horikawa et al., 2007; Martin & Goodnow, 2002). On a cellular level, an antigen-mediated engagement of the IgG BCR results in a lower level of cell death in the highly proliferating antigen-specific B cells (Martin & Goodnow, 2002).

IgG-expressing lymphocytes show a higher plasma-cell differentiation capacity. The activation of the IgG BCR results in a higher level of the intracellular Ca^{2+} as compared to IgM, and it is specifically supported by the intracellular tail of the γHC (Horikawa et al., 2007). Intriguingly, apart from the stronger primary Ca flux, the engagement of the IgG BCR does not lead to significant change in the expression of the Ca/calmodulin-dependent set of genes, neither does it result in stronger Erk and Akt pathways activation (Horikawa et al., 2007; Waisman et al., 2007). Several groups investigated involvement of the CD22 inhibitory receptor in the extent of BCR-mediated activation. The study on K46 cell line provided biochemical evidence that, in contrast to IgM, antigen-mediated IgG stimulation is resistant to the inhibitory activity of the CD22 on BCR signaling (Wakabayashi, Adachi, Wienands, & Tsubata, 2002). Yet convincing experimental evidence supported the opposite conclusion, reported in two studies. It was shown in two independent *in vivo* systems that enhanced activation response of IgG BCR is not uncoupled from the CD22 involvement and should originate from other mechanisms (Horikawa et al., 2007; Waisman et al., 2007).

A stronger support of the plasma cell differentiation is not exquisitely conferred by the IgG type of BCR itself. Examining the plasma cell responses upon stimulation of naive IgG-expressing B cells and antigen-experienced cells revealed that the responsiveness strongly depends on the previous history of the B cell. Likewise, Ag-experienced ("true" memory) cells are much more potent in effector plasma cell differentiation compared with the naive cells with the same IgG BCR (Kometani et al., 2013).

In spite of bearing an intracellular tail, IgG receptors are not independent of the Ig-α/Ig-β signal-transducing heterodimer, yet rudimentary B-cell development can be supported by the expression of the IgG1 BCR in combination with tailless Ig-α (Waisman et al., 2007). In the periphery, the expression of the IgG BCR results in a longer half-life of the B cells (Waisman et al., 2007).

The expression of the IgG1 from the early development results in a mild block at the pre-B cell stage (Dougan et al., 2012; Waisman et al., 2007) and decreased support for the differentiation of B1 B cells. This phenotype corresponds to a mild block in B-cell development observed in the IgM knockout mice (Lutz et al., 1998; Ubelhart et al., 2010), where development is supported exclusively by the expression of the δHC-bearing receptors.

6. IMPORTANCE OF BCR ISOTYPE IN B-CELL LYMPHOMAS

BCR plays an essential role in the maintenance of mature B cells. Even in the absence of the specific ligand, tonic BCR signaling supports B-cell survival (Kraus et al., 2004; Lam et al., 1997; Srinivasan et al., 2009). Therefore, it is not surprising that mutations accompanying malignant B-cell transformation often feed into the BCR signaling network (Rickert, 2013; Young & Staudt, 2013). The role of a ligand-dependent or ligand-independent BCR signaling in malignant B cells is often not resolved, yet many of the lymphomas of mature B cells carry footprints of passing through the germinal center (GC) reaction—diversification of the IgH gene locus via somatic hypermutation (SHM) and class-switch recombination. Surprisingly, the presence of the mutations in the variable Ig sequences, indicative of SHM, is often not complemented with class-switch recombination in the constant region of the IgH. Preference for IgM receptors is characteristic for activated B-cell-like diffuse large B-cell lymphoma (ABC DLBCL), Burkitt's lymphoma (BL), follicular lymphoma (FL), chronic lymphocytic leukemia (CLL), mantle cell lymphoma (MCL), and mucosa-associated lymphoid tissue lymphoma/marginal zone lymphoma (MALT/MZL; Young & Staudt, 2013).

The following experimental evidence supports a selective pressure for IgM expression in case of follicular lymphoma, also known as "allelic paradox." FL cells carry a characteristic *t*(14;18) chromosomal translocation, which brings together the bcl-2 gene and IgH enhancer. *t*(14;18) translocation disrupts the variable part of the respective IgH allele, and therefore, translocated IgH gene locus cannot produce an HC polypeptide. The second copy of the IgH allele in FL cells always carries a productive VDJ rearrangement and encodes a functional HC. In the majority of cases, the productive IgH allele retains the μHC coding region, while the translocated allele often undergoes class switching to IgG (Vaandrager et al., 1998). Moreover, in some cases, retention of the μHC region was even accompanied by deletions in the 3′ part of the IgH gene locus, resulting from recombination between the downstream switch regions (Vaandrager et al., 1998).

Interestingly, retention of the IgM BCR was also reported in the mouse models of FL (Egle, Harris, Bath, O'Reilly, & Cory, 2004; McDonnell et al., 1989).

Molecular hints, suggestive of selective retention of the IgM coding region in the IgH gene locus, were as well detected in human ABC DLBCL. Analysis of 31 ABC DLBCL samples revealed that compared to the other subtypes of DLBCL malignancies, this type of cancer is characterized by the presence of illegitimate class-switch recombination events (Lenz et al., 2007). Almost one-third of the cases are characterized by aberrant mutations within the switch element Sμ, including deletions and insertions that occur of the productive IgH allele. The nonproductive IgH allele in ABC DLBCL samples often lacks the μHC coding region. In contrast to that, majority of the germinal center B-cell-like diffuse large B-cell lymphoma (GCB DLBCL) cases are characterized by the legitimate class-switch recombination of IgH gene locus and the expression of other than IgM BCR isotypes (Ruminy et al., 2011). Apart from the distinct BCR isotypes, ABC and GCB DLBCL show disparate gene-expression pattern and contrasting dependence on BCR signaling. A number of genetic aberrations specifically suggest that ABC DLBCL cells are particularly dependent on the BCR proximal signaling. For example, recurrent mutations in the ITAM motifs of Ig-α of Ig-β, often found in the ABC DLBCL, support higher BCR expression on the cell surface (Davis et al., 2010). Furthermore, siRNA-mediated restriction of the IgM, Igκ, or Ig-α expression in ABC DLBCL cells results in prominent decline of lymphoma cell survival, while GCB DLBCLs are mostly resistant to the same BCR targeting (Davis et al., 2010).

In addition to ABC DLBCL, genetic lesions affecting the Sμ switch region (and therefore preventing its use for class-switch recombination) were also found in primary lymphomas of the central nervous system (Montesinos-Rongen et al., 2005) as well as in CLL (Nardini et al., 2002). Presented molecular signatures of selective retention of the IgM expression by the wide number of human B-cell lymphomas stimulate to think of a specific role of IgM BCR isotype in supporting pathogenesis of such malignant cells. Therefore, a detailed understanding of BCR isotype-specific functions of IgM and IgG holds a promise to advance the therapeutic strategies for efficient lymphoma targeting. Our previous work on CLL receptors revealed that in contrast to many other lymphomas, CLL BCRs show cell-autonomous receptor activity in the receptor reconstitution system (Duhren-von Minden et al., 2012). Moreover, our unpublished results on high IgM sensitivity to the presence of autoantigens (E. Surova and H. Jumaa, manuscript in preparation), supported by a number of previous studies, can provide a possible explanation to a selective use of IgM in a number of B-cell lymphomas. A low threshold of IgM BCR activation (possibly by autoantigens or due to specific BCR conformations) results in sustained signaling, thereby contributing to a competitive survival advantage of malignant B cells expressing IgM.

7. CONCLUDING REMARKS

The functions of B cells in orchestrating immune reaction are of great importance to basic and clinical research. Many studies over the recent years demonstrated that different isotypes of B-cell antigen receptor as well as soluble antibodies possess distinct properties in inducing B-cell activation upon interaction with the cognate antigen. Not all molecular determinants of distinct types of B-cell responses depending on the BCR isotype are substantially understood. Yet the work of many groups convincingly demonstrated the critical function of the μHC in the context of the precursor and conventional IgM B-cell receptor in early B-cell development. Several lines of evidence support the important, previously underappreciated, role of self-reactivity during early B-cell development. Interestingly, continuous exposure of the B cells to autoantigens might hold a promise to understand how mature B cells maintain in the periphery and regulate responsiveness to self-antigens and non-self-antigens. A number of studies provided evidence for the important role of IgD in peripheral tolerance.

Moreover, considerable progress has been achieved in understanding the signaling of IgG BCR. Specific features of IgA and IgE immunoglobulin receptors activation and downstream signaling remain largely undisclosed. Yet, the relevance of IgA and IgE to the maintenance/break of complex immune homeostasis at the mucosal surfaces and skin is of critical importance to basic research and clinical practice.

ACKNOWLEDGMENTS

We thank Dr. Peter Nielsen and Philipp Rauf for their comments on the draft of the manuscript. The authors were supported by the Deutsche Forschungsgemeinschaft (grants JU463/2-1 to H. J. and GSC-4 to Spemann Graduate School) and Deutsche Krebshilfe (Project 108935).

REFERENCES

Achatz, G., Nitschke, L., & Lamers, M. C. (1997). Effect of transmembrane and cytoplasmic domains of IgE on the IgE response. *Science, 276*(5311), 409–411.

Ait-Azzouzene, D., Verkoczy, L., Peters, J., Gavin, A., Skog, P., Vela, J. L., et al. (2005). An immunoglobulin C kappa-reactive single chain antibody fusion protein induces tolerance through receptor editing in a normal polyclonal immune system. *The Journal of Experimental Medicine, 201*(5), 817–828.

Arnold, L. W., Pennell, C. A., McCray, S. K., & Clarke, S. H. (1994). Development of B-1 cells: Segregation of phosphatidyl choline-specific B cells to the B-1 population occurs after immunoglobulin gene expression. *The Journal of Experimental Medicine, 179*(5), 1585–1595.

Arnold, J. N., Wormald, M. R., Sim, R. B., Rudd, P. M., & Dwek, R. A. (2007). The impact of glycosylation on the biological function and structure of human immunoglobulins. *Annual Review of Immunology, 25*, 21–50.

Bankovich, A. J., Raunser, S., Juo, Z. S., Walz, T., Davis, M. M., & Garcia, K. C. (2007). Structural insight into pre-B cell receptor function. *Science, 316*(5822), 291–294.

Billingham, R. E., Brent, L., & Medawar, P. B. (1956). The antigenic stimulus in transplantation immunity. *Nature, 178*(4532), 514–519.

Bradl, H., Wittmann, J., Milius, D., Vettermann, C., & Jack, H. M. (2003). Interaction of murine precursor B cell receptor with stroma cells is controlled by the unique tail of lambda 5 and stroma cell-associated heparan sulfate. *The Journal of Immunology, 171*(5), 2338–2348.

Brink, R., Goodnow, C. C., & Basten, A. (1995). IgD expression on B cells is more efficient than IgM but both receptors are functionally equivalent in up-regulation CD80/CD86 co-stimulatory molecules. *European Journal of Immunology, 25*(7), 1980–1984.

Brink, R., Goodnow, C. C., Crosbie, J., Adams, E., Eris, J., Mason, D. Y., et al. (1992). Immunoglobulin M and D antigen receptors are both capable of mediating B lymphocyte activation, deletion, or anergy after interaction with specific antigen. *The Journal of Experimental Medicine, 176*(4), 991–1005.

Burnet, F. M. (1959). *Clonal selection theory of acquired immunity*. London: Cambridge University Press.

Cambier, J. C., Gauld, S. B., Merrell, K. T., & Vilen, B. J. (2007). B-cell anergy: From transgenic models to naturally occurring anergic B cells? *Nature Reviews. Immunology, 7*(8), 633–643.

Carsetti, R., Kohler, G., & Lamers, M. C. (1993). A role for immunoglobulin D: Interference with tolerance induction. *European Journal of Immunology, 23*(1), 168–178.

Casellas, R., Shih, T. A., Kleinewietfeld, M., Rakonjac, J., Nemazee, D., Rajewsky, K., et al. (2001). Contribution of receptor editing to the antibody repertoire. *Science, 291*(5508), 1541–1544.

Casola, S., Otipoby, K. L., Alimzhanov, M., Humme, S., Uyttersprot, N., Kutok, J. L., et al. (2004). B cell receptor signal strength determines B cell fate. *Nature Immunology, 5*(3), 317–327.

Cerny, A., Heusser, C., Sutter, S., Huegin, A. W., Bazin, H., Hengartner, H., et al. (1986). Generation of agammaglobulinaemic mice by prenatal and postnatal exposure to polyclonal or monoclonal anti-IgM antibodies. *Scandinavian Journal of Immunology, 24*(4), 437–445.

Chen, C., Nagy, Z., Prak, E. L., & Weigert, M. (1995). Immunoglobulin heavy chain gene replacement: A mechanism of receptor editing. *Immunity, 3*(6), 747–755.

Cheng, A. M., Rowley, B., Pao, W., Hayday, A., Bolen, J. B., & Pawson, T. (1995). Syk tyrosine kinase required for mouse viability and B-cell development. *Nature, 378*(6554), 303–306.

Choi, N. M., Loguercio, S., Verma-Gaur, J., Degner, S. C., Torkamani, A., Su, A. I., et al. (2013). Deep sequencing of the murine IgH repertoire reveals complex regulation of nonrandom v gene rearrangement frequencies. *The Journal of Immunology, 191*(5), 2393–2402.

Cooper, L. J., Robertson, D., Granzow, R., & Greenspan, N. S. (1994). Variable domain-identical antibodies exhibit IgG subclass-related differences in affinity and kinetic constants as determined by surface plasmon resonance. *Molecular Immunology, 31*(8), 577–584.

Corper, A. L., Sohi, M. K., Bonagura, V. R., Steinitz, M., Jefferis, R., Feinstein, A., et al. (1997). Structure of human IgM rheumatoid factor Fab bound to its autoantigen IgG Fc reveals a novel topology of antibody-antigen interaction. *Nature Structural Biology, 4*(5), 374–381.

Dalziel, M., Crispin, M., Scanlan, C. N., Zitzmann, N., & Dwek, R. A. (2014). Emerging principles for the therapeutic exploitation of glycosylation. *Science, 343*(6166), 37–45.

Davis, R. E., Ngo, V. N., Lenz, G., Tolar, P., Young, R. M., Romesser, P. B., et al. (2010). Chronic active B-cell-receptor signalling in diffuse large B-cell lymphoma. *Nature, 463*(7277), 88–92.

De Genst, E., Saerens, D., Muyldermans, S., & Conrath, K. (2006). Antibody repertoire development in camelids. *Developmental and Comparative Immunology, 30*(1–2), 187–198.

Dougan, S. K., Ogata, S., Hu, C. C., Grotenbreg, G. M., Guillen, E., Jaenisch, R., et al. (2012). IgG1 + ovalbumin-specific B-cell transnuclear mice show class switch recombination in rare allelically included B cells. *Proceedings of the National Academy of Sciences of the United States of America, 109*(34), 13739–13744 [Research Support, N.I.H., Extramural Research Support, Non-U.S. Gov't].

Doyle-Cooper, C., Hudson, K. E., Cooper, A. B., Ota, T., Skog, P., Dawson, P. E., et al. (2013). Immune tolerance negatively regulates B cells in knock-in mice expressing broadly neutralizing HIV antibody 4E10. *The Journal of Immunology, 191*(6), 3186–3191.

Duhren-von Minden, M., Ubelhart, R., Schneider, D., Wossning, T., Bach, M. P., Buchner, M., et al. (2012). Chronic lymphocytic leukaemia is driven by antigen-independent cell-autonomous signalling. *Nature, 489*(7415), 309–312.

Edholm, E. S., Bengten, E., & Wilson, M. (2011). Insights into the function of IgD. *Developmental and Comparative Immunology, 35*(12), 1309–1316.

Egle, A., Harris, A. W., Bath, M. L., O'Reilly, L., & Cory, S. (2004). VavP-Bcl2 transgenic mice develop follicular lymphoma preceded by germinal center hyperplasia. *Blood, 103*(6), 2276–2283.

Elantak, L., Espeli, M., Boned, A., Bornet, O., Bonzi, J., Gauthier, L., et al. (2012). Structural basis for galectin-1-dependent pre-B cell receptor (pre-BCR) activation. *The Journal of Biological Chemistry, 287*(53), 44703–44713.

Engels, N., Konig, L. M., Heemann, C., Lutz, J., Tsubata, T., Griep, S., et al. (2009). Recruitment of the cytoplasmic adaptor Grb2 to surface IgG and IgE provides antigen receptor-intrinsic costimulation to class-switched B cells. *Nature Immunology, 10*(9), 1018–1025. http://dx.doi.org/10.1038/ni.1764.

Erikson, J., Radic, M. Z., Camper, S. A., Hardy, R. R., Carmack, C., & Weigert, M. (1991). Expression of anti-DNA immunoglobulin transgenes in non-autoimmune mice. *Nature, 349*(6307), 331–334.

Eschbach, C., Bach, M. P., Fidler, I., Pelanda, R., Kohler, F., Rajewsky, K., et al. (2011). Efficient generation of B lymphocytes by recognition of self-antigens. *European Journal of Immunology, 41*(8), 2397–2403.

Espeli, M., Mancini, S. J., Breton, C., Poirier, F., & Schiff, C. (2009). Impaired B-cell development at the pre-BII-cell stage in galectin-1-deficient mice due to inefficient pre-BII/stromal cell interactions. *Blood, 113*(23), 5878–5886.

Finkelman, F. D., Mond, J. J., & Metcalf, E. S. (1983). Effects of neonatal anti-delta antibody treatment on the murine immune system. I. Suppression of development of surface IgD + B cells and expansion of a surface IgM + IgD- B lymphocyte population. *The Journal of Immunology, 131*(2), 593–600.

Flajnik, M. F. (2002). Comparative analyses of immunoglobulin genes: Surprises and portents. *Nature Reviews. Immunology, 2*(9), 688–698.

Galler, G. R., Mundt, C., Parker, M., Pelanda, R., Martensson, I. L., & Winkler, T. H. (2004). Surface mu heavy chain signals down-regulation of the V(D)J-recombinase machinery in the absence of surrogate light chain components. *The Journal of Experimental Medicine, 199*(11), 1523–1532.

Gauthier, L., Rossi, B., Roux, F., Termine, E., & Schiff, C. (2002). Galectin-1 is a stromal cell ligand of the pre-B cell receptor (BCR) implicated in synapse formation between pre-B and stromal cells and in pre-BCR triggering. *Proceedings of the National Academy of Sciences of the United States of America, 99*(20), 13014–13019.

Gay, D., Saunders, T., Camper, S., & Weigert, M. (1993). Receptor editing: An approach by autoreactive B cells to escape tolerance. *The Journal of Experimental Medicine, 177*(4), 999–1008.

Goodnow, C. C., Crosbie, J., Adelstein, S., Lavoie, T. B., Smith-Gill, S. J., Brink, R. A., et al. (1988). Altered immunoglobulin expression and functional silencing of self-reactive B lymphocytes in transgenic mice. *Nature, 334*(6184), 676–682.

Gordon, J., Murgita, R. A., & Tomasi, T. B., Jr. (1975). The immune response of mice treated with anti-mu antibodies: The effect on antibody-forming cells, their precursors and helper cells assayed in vitro. *The Journal of Immunology, 114*(6), 1808–1812.

Grandien, A., Fucs, R., Nobrega, A., Andersson, J., & Coutinho, A. (1994). Negative selection of multireactive B cell clones in normal adult mice. *European Journal of Immunology, 24*(6), 1345–1352.

Grawunder, U., Leu, T. M., Schatz, D. G., Werner, A., Rolink, A. G., Melchers, F., et al. (1995). Down-regulation of RAG1 and RAG2 gene expression in preB cells after functional immunoglobulin heavy chain rearrangement. *Immunity, 3*(5), 601–608.

Greenspan, N. S., & Cooper, L. J. (1993). Cooperative binding by mouse IgG3 antibodies: Implications for functional affinity, effector function, and isotype restriction. *Springer Seminars in Immunopathology, 15*(2–3), 275–291.

Harwood, N. E., & Batista, F. D. (2010). Early Events in B Cell Activation. *Annual Review of Immunology, 28*(1), 185–210.

Hasan, M., Polic, B., Bralic, M., Jonjic, S., & Rajewsky, K. (2002). Incomplete block of B cell development and immunoglobulin production in mice carrying the μ MT mutation on the BALB/c background. *European Journal of Immunology, 32*(12), 3463–3471.

Hauser, J., Sveshnikova, N., Wallenius, A., Baradaran, S., Saarikettu, J., & Grundstrom, T. (2008). B-cell receptor activation inhibits AID expression through calmodulin inhibition of E-proteins. *Proceedings of the National Academy of Sciences of the United States of America, 105*(4), 1267–1272.

Hayakawa, K., Asano, M., Shinton, S. A., Gui, M., Allman, D., Stewart, C. L., et al. (1999). Positive selection of natural autoreactive B cells. *Science, 285*(5424), 113–116.

Herzog, S., Hug, E., Meixlsperger, S., Paik, J. H., DePinho, R. A., Reth, M., et al. (2008). SLP-65 regulates immunoglobulin light chain gene recombination through the PI(3)K-PKB-Foxo pathway. *Nature Immunology, 9*(6), 623–631.

Herzog, S., Reth, M., & Jumaa, H. (2009). Regulation of B-cell proliferation and differentiation by pre-B-cell receptor signalling. *Nature Reviews. Immunology, 9*(3), 195–205.

Hess, J., Werner, A., Wirth, T., Melchers, F., Jack, H. M., & Winkler, T. H. (2001). Induction of pre-B cell proliferation after de novo synthesis of the pre-B cell receptor. *Proceedings of the National Academy of Sciences of the United States of America, 98*(4), 1745–1750.

Hippen, K. L., Schram, B. R., Tze, L. E., Pape, K. A., Jenkins, M. K., & Behrens, T. W. (2005). In vivo assessment of the relative contributions of deletion, anergy, and editing to B cell self-tolerance. *The Journal of Immunology, 175*(2), 909–916.
Horikawa, K., Martin, S. W., Pogue, S. L., Silver, K., Peng, K., Takatsu, K., et al. (2007). Enhancement and suppression of signaling by the conserved tail of IgG memory-type B cell antigen receptors. *The Journal of Experimental Medicine, 204*(4), 759–769.
Hug, E., Hobeika, E., Reth, M., & Jumaa, H. (2013). Inducible expression of hyperactive Syk in B cells activates Blimp-1-dependent terminal differentiation. *Oncogene*. http://dx.doi.org/10.1038/onc.2013.326.
James, L. C., Roversi, P., & Tawfik, D. S. (2003). Antibody multispecificity mediated by conformational diversity. *Science, 299*(5611), 1362–1367.
Janssens, R., Dekker, S., Hendriks, R. W., Panayotou, G., van Remoortere, A., San, J. K., et al. (2006). Generation of heavy-chain-only antibodies in mice. *Proceedings of the National Academy of Sciences of the United States of America, 103*(41), 15130–15135.
Jiang, H., Chang, F. C., Ross, A. E., Lee, J., Nakayama, K., & Desiderio, S. (2005). Ubiquitylation of RAG-2 by Skp2-SCF links destruction of the V(D)J recombinase to the cell cycle. *Molecular Cell, 18*(6), 699–709.
Johnston, C. M., Wood, A. L., Bolland, D. J., & Corcoran, A. E. (2006). Complete sequence assembly and characterization of the C57BL/6 mouse Ig heavy chain V region. *The Journal of Immunology, 176*(7), 4221–4234.
Jumaa, H., Wollscheid, B., Mitterer, M., Wienands, J., Reth, M., & Nielsen, P. J. (1999). Abnormal development and function of B lymphocytes in mice deficient for the signaling adaptor protein SLP-65. *Immunity, 11*(5), 547–554.
Kaisho, T., Schwenk, F., & Rajewsky, K. (1997). The roles of gamma 1 heavy chain membrane expression and cytoplasmic tail in IgG1 responses. *Science, 276*(5311), 412–415.
Karasuyama, H., Kudo, A., & Melchers, F. (1990). The proteins encoded by the VpreB and lambda 5 pre-B cell-specific genes can associate with each other and with mu heavy chain. *The Journal of Experimental Medicine, 172*(3), 969–972.
Karasuyama, H., Rolink, A., & Melchers, F. (1993). A complex of glycoproteins is associated with VpreB/lambda 5 surrogate light chain on the surface of mu heavy chain-negative early precursor B cell lines. *The Journal of Experimental Medicine, 178*(2), 469–478.
Karasuyama, H., Rolink, A., Shinkai, Y., Young, F., Alt, F. W., & Melchers, F. (1994). The expression of Vpre-B/lambda 5 surrogate light chain in early bone marrow precursor B cells of normal and B cell-deficient mutant mice. *Cell, 77*(1), 133–143.
Keenan, R. A., De Riva, A., Corleis, B., Hepburn, L., Licence, S., Winkler, T. H., et al. (2008). Censoring of autoreactive B cell development by the pre-B cell receptor. *Science, 321*(5889), 696–699.
Keyna, U., Applequist, S. E., Jongstra, J., Beck-Engeser, G. B., & Jack, H. M. (1995). Ig mu heavy chains with VH81X variable regions do not associate with lambda 5. *Annals of the New York Academy of Sciences, 764*, 39–42.
Keyna, U., Beck-Engeser, G. B., Jongstra, J., Applequist, S. E., & Jack, H. M. (1995). Surrogate light chain-dependent selection of Ig heavy chain V regions. *The Journal of Immunology, 155*(12), 5536–5542.
Kim, K. M., & Reth, M. (1995). The B cell antigen receptor of class IgD induces a stronger and more prolonged protein tyrosine phosphorylation than that of class IgM. *The Journal of Experimental Medicine, 181*(3), 1005–1014.
Kitamura, D., Kudo, A., Schaal, S., Muller, W., Melchers, F., & Rajewsky, K. (1992). A critical role of lambda 5 protein in B cell development. *Cell, 69*(5), 823–831.
Kitamura, D., & Rajewsky, K. (1992). Targeted disruption of mu chain membrane exon causes loss of heavy-chain allelic exclusion. *Nature, 356*(6365), 154–156.

Koelsch, K., Zheng, N. Y., Zhang, Q., Duty, A., Helms, C., Mathias, M. D., et al. (2007). Mature B cells class switched to IgD are autoreactive in healthy individuals. *The Journal of Clinical Investigation*, *117*(6), 1558–1565.

Kohler, F., Hug, E., Eschbach, C., Meixlsperger, S., Hobeika, E., Kofer, J., et al. (2008). Autoreactive B cell receptors mimic autonomous pre-B cell receptor signaling and induce proliferation of early B cells. *Immunity*, *29*(6), 912–921.

Kometani, K., Nakagawa, R., Shinnakasu, R., Kaji, T., Rybouchkin, A., Moriyama, S., et al. (2013). Repression of the transcription factor Bach2 contributes to predisposition of IgG1 memory B cells toward plasma cell differentiation. *Immunity*, *39*(1), 136–147.

Kraus, M., Alimzhanov, M. B., Rajewsky, N., & Rajewsky, K. (2004). Survival of resting mature B lymphocytes depends on BCR signaling via the Igalpha/beta heterodimer. *Cell*, *117*(6), 787–800.

Kreslavsky, T., Kim, H.-J., Koralov, S. B., Ghitza, D., Buch, T., Cantor, H., et al. (2013). Negative selection, not receptor editing, is a physiological response of autoreactive thymocytes. *The Journal of Experimental Medicine*, *210*(10), 1911–1918.

Kurosaki, T. (2011). Regulation of BCR signaling. *Molecular Immunology*, *48*(11), 1287–1291.

Lam, K. P., Kuhn, R., & Rajewsky, K. (1997). In vivo ablation of surface immunoglobulin on mature B cells by inducible gene targeting results in rapid cell death. *Cell*, *90*(6), 1073–1083.

Lang, J., Jackson, M., Teyton, L., Brunmark, A., Kane, K., & Nemazee, D. (1996). B cells are exquisitely sensitive to central tolerance and receptor editing induced by ultralow affinity, membrane-bound antigen. *The Journal of Experimental Medicine*, *184*(5), 1685–1697.

Lawton, A. R., 3rd, Asofsky, R., Hylton, M. B., & Cooper, M. D. (1972). Suppression of immunoglobulin class synthesis in mice. I. Effects of treatment with antibody to -chain. *The Journal of Experimental Medicine*, *135*(2), 277–297.

Lenz, G., Nagel, I., Siebert, R., Roschke, A. V., Sanger, W., Wright, G. W., et al. (2007). Aberrant immunoglobulin class switch recombination and switch translocations in activated B cell-like diffuse large B cell lymphoma. *The Journal of Experimental Medicine*, *204*(3), 633–643.

Li, Z., Dordai, D. I., Lee, J., & Desiderio, S. (1996). A conserved degradation signal regulates RAG-2 accumulation during cell division and links V(D)J recombination to the cell cycle. *Immunity*, *5*(6), 575–589.

Lin, W. C., & Desiderio, S. (1994). Cell cycle regulation of V(D)J recombination-activating protein RAG-2. *Proceedings of the National Academy of Sciences of the United States of America*, *91*(7), 2733–2737.

Liu, F., Bergami, P. L., Duval, M., Kuhrt, D., Posner, M., & Cavacini, L. (2003). Expression and functional activity of isotype and subclass switched human monoclonal antibody reactive with the base of the V3 loop of HIV-1 gp120. *AIDS Research and Human Retroviruses*, *19*(7), 597–607.

Liu, H., & May, K. (2012). Disulfide bond structures of IgG molecules: Structural variations, chemical modifications and possible impacts to stability and biological function. *MAbs*, *4*(1), 17–23.

Lutz, C., Ledermann, B., Kosco-Vilbois, M. H., Ochsenbein, A. F., Zinkernagel, R. M., Kohler, G., et al. (1998). IgD can largely substitute for loss of IgM function in B cells. *Nature*, *393*(6687), 797–801.

Macpherson, A. J., Lamarre, A., McCoy, K., Harriman, G. R., Odermatt, B., Dougan, G., et al. (2001). IgA production without mu or delta chain expression in developing B cells. *Nature Immunology*, *2*(7), 625–631.

Magadan-Mompo, S., Sanchez-Espinel, C., & Gambon-Deza, F. (2011). Immunoglobulin heavy chains in medaka (Oryzias latipes). *BMC Evolutionary Biology*, *11*, 165.

Manivel, V., Sahoo, N. C., Salunke, D. M., & Rao, K. V. (2000). Maturation of an antibody response is governed by modulations in flexibility of the antigen-combining site. *Immunity*, *13*(5), 611–620.

Manning, D. D., & Jutila, J. W. (1972). Immunosuppression of mice injected with heterologous anti-immunoglobulin heavy chain antisera. *The Journal of Experimental Medicine*, *135*(6), 1316–1333.
Martensson, A., Argon, Y., Melchers, F., Dul, J. L., & Martensson, I. L. (1999). Partial block in B lymphocyte development at the transition into the pre-B cell receptor stage in Vpre-B1-deficient mice. *International Immunology*, *11*(3), 453–460.
Martin, D. A., Bradl, H., Collins, T. J., Roth, E., Jack, H. M., & Wu, G. E. (2003). Selection of Ig mu heavy chains by complementarity-determining region 3 length and amino acid composition. *The Journal of Immunology*, *171*(9), 4663–4671.
Martin, S. W., & Goodnow, C. C. (2002). Burst-enhancing role of the IgG membrane tail as a molecular determinant of memory. *Nature Immunology*, *3*(2), 182–188.
Mattila, P. K., Feest, C., Depoil, D., Treanor, B., Montaner, B., Otipoby, K. L., et al. (2013). The actin and tetraspanin networks organize receptor nanoclusters to regulate B cell receptor-mediated signaling. *Immunity*, *38*(3), 461–474.
McDonnell, T. J., Deane, N., Platt, F. M., Nunez, G., Jaeger, U., McKearn, J. P., et al. (1989). bcl-2-immunoglobulin transgenic mice demonstrate extended B cell survival and follicular lymphoproliferation. *Cell*, *57*(1), 79–88.
Meixlsperger, S., Kohler, F., Wossning, T., Reppel, M., Muschen, M., & Jumaa, H. (2007). Conventional light chains inhibit the autonomous signaling capacity of the B cell receptor. *Immunity*, *26*(3), 323–333.
Melamed, D., Miri, E., Leider, N., & Nemazee, D. (2000). Unexpected autoantibody production in membrane Ig-mu-deficient/lpr mice. *The Journal of Immunology*, *165*(8), 4353–4358.
Melchers, F. (2005). The pre-B-cell receptor: Selector of fitting immunoglobulin heavy chains for the B-cell repertoire. *Nature Reviews. Immunology*, *5*(7), 578–584.
Meng, W., Yunk, L., Wang, L. S., Maganty, A., Xue, E., Cohen, P. L., et al. (2011). Selection of individual VH genes occurs at the pro-B to pre-B cell transition. *The Journal of Immunology*, *187*(4), 1835–1844.
Menoret, S., Iscache, A. L., Tesson, L., Remy, S., Usal, C., Osborn, M. J., et al. (2010). Characterization of immunoglobulin heavy chain knockout rats. *European Journal of Immunology*, *40*(10), 2932–2941.
Minegishi, Y., & Conley, M. E. (2001). Negative selection at the pre-BCR checkpoint elicited by human mu heavy chains with unusual CDR3 regions. *Immunity*, *14*(5), 631–641.
Minegishi, Y., Coustan-Smith, E., Wang, Y. H., Cooper, M. D., Campana, D., & Conley, M. E. (1998). Mutations in the human lambda5/14.1 gene result in B cell deficiency and agammaglobulinemia. *The Journal of Experimental Medicine*, *187*(1), 71–77.
Minegishi, Y., Hendershot, L. M., & Conley, M. E. (1999). Novel mechanisms control the folding and assembly of lambda5/14.1 and VpreB to produce an intact surrogate light chain. *Proceedings of the National Academy of Sciences of the United States of America*, *96*(6), 3041–3046.
Misener, V., Downey, G. P., & Jongstra, J. (1991). The immunoglobulin light chain related protein lambda 5 is expressed on the surface of mouse pre-B cell lines and can function as a signal transducing molecule. *International Immunology*, *3*(11), 1129–1136.
Montesinos-Rongen, M., Schmitz, R., Courts, C., Stenzel, W., Bechtel, D., Niedobitek, G., et al. (2005). Absence of immunoglobulin class switch in primary lymphomas of the central nervous system. *The American Journal of Pathology*, *166*(6), 1773–1779.
Morahan, G., Berek, C., & Miller, J. F. (1983). An idiotypic determinant formed by both immunoglobulin constant and variable regions. *Nature*, *301*(5902), 720–722.
Morelock, M. M., Rothlein, R., Bright, S. M., Robinson, M. K., Graham, E. T., Sabo, J. P., et al. (1994). Isotype choice for chimeric antibodies affects binding properties. *The Journal of Biological Chemistry*, *269*(17), 13048–13055.

Mundt, C., Licence, S., Shimizu, T., Melchers, F., & Martensson, I. L. (2001). Loss of precursor B cell expansion but not allelic exclusion in VpreB1/VpreB2 double-deficient mice. *The Journal of Experimental Medicine*, *193*(4), 435–445.

Nagata, K., Nakamura, T., Kitamura, F., Kuramochi, S., Taki, S., Campbell, K. S., et al. (1997). The Ig alpha/Igbeta heterodimer on mu-negative proB cells is competent for transducing signals to induce early B cell differentiation. *Immunity*, 7(4), 559–570.

Nardini, E., Rizzi, S., Capello, D., Vitolo, U., Gaidano, G., Menard, S., et al. (2002). Most immunoglobulin heavy chain switch mu rearrangements in B-cell chronic lymphocytic leukemia are internal deletions. *FEBS Letters*, *518*(1–3), 119–123.

Nemazee, D., & Buerki, K. (1989). Clonal deletion of autoreactive B lymphocytes in bone marrow chimeras. *Proceedings of the National Academy of Sciences of the United States of America*, *86*(20), 8039–8043.

Nemazee, D. A., & Burki, K. (1989). Clonal deletion of B lymphocytes in a transgenic mouse bearing anti-MHC class I antibody genes. *Nature*, *337*(6207), 562–566.

Nemazee, D., & Weigert, M. (2000). Revising B cell receptors. *The Journal of Experimental Medicine*, *191*(11), 1813–1817.

Nitschke, L., Kosco, M. H., Kohler, G., & Lamers, M. C. (1993). Immunoglobulin D-deficient mice can mount normal immune responses to thymus-independent and -dependent antigens. *Proceedings of the National Academy of Sciences of the United States of America*, *90*(5), 1887–1891.

Novobrantseva, T. I., Martin, V. M., Pelanda, R., Muller, W., Rajewsky, K., & Ehlich, A. (1999). Rearrangement and expression of immunoglobulin light chain genes can precede heavy chain expression during normal B cell development in mice. *The Journal of Experimental Medicine*, *189*(1), 75–88.

Novobrantseva, T., Xu, S., Tan, J. E.-L., Maruyama, M., Schwers, S., Pelanda, R., et al. (2005). Stochastic pairing of Ig heavy and light chains frequently generates B cell antigen receptors that are subject to editing in vivo. *International Immunology*, *17*(4), 343–350.

Ohnishi, K., & Melchers, F. (2003). The nonimmunoglobulin portion of lambda5 mediates cell-autonomous pre-B cell receptor signaling. *Nature Immunology*, *4*(9), 849–856.

Ohnishi, K., Melchers, F., & Shimizu, T. (2005). Lymphocyte-expressed BILL-cadherin/cadherin-17 contributes to the development of B cells at two stages. *European Journal of Immunology*, *35*(3), 957–963.

Ohnishi, K., Shimizu, T., Karasuyama, H., & Melchers, F. (2000). The identification of a nonclassical cadherin expressed during B cell development and its interaction with surrogate light chain. *The Journal of Biological Chemistry*, *275*(40), 31134–31144.

Ohta, Y., & Flajnik, M. (2006). IgD, like IgM, is a primordial immunoglobulin class perpetuated in most jawed vertebrates. *Proceedings of the National Academy of Sciences of the United States of America*, *103*(28), 10723–10728.

Okkenhaug, K., Bilancio, A., Farjot, G., Priddle, H., Sancho, S., Peskett, E., et al. (2002). Impaired B and T cell antigen receptor signaling in p110delta PI 3-kinase mutant mice. *Science*, *297*(5583), 1031–1034.

Ozato, K., Mayer, N., & Sachs, D. H. (1980). Hybridoma cell lines secreting monoclonal antibodies to mouse H-2 and Ia antigens. *The Journal of Immunology*, *124*(2), 533–540.

Papavasiliou, F., Misulovin, Z., Suh, H., & Nussenzweig, M. C. (1995). The role of Ig beta in precursor B cell transition and allelic exclusion. *Science*, *268*(5209), 408–411.

Parker, M. J., Licence, S., Erlandsson, L., Galler, G. R., Chakalova, L., Osborne, C. S., et al. (2005). The pre-B-cell receptor induces silencing of VpreB and lambda5 transcription. *The EMBO Journal*, *24*(22), 3895–3905.

Pelanda, R., Braun, U., Hobeika, E., Nussenzweig, M. C., & Reth, M. (2002). B cell progenitors are arrested in maturation but have intact VDJ recombination in the absence of Ig-alpha and Ig-beta. *The Journal of Immunology*, *169*(2), 865–872.

Pelanda, R., Schaal, S., Torres, R. M., & Rajewsky, K. (1996). A prematurely expressed Ig(kappa) transgene, but not V(kappa)J(kappa) gene segment targeted into the Ig(kappa) locus, can rescue B cell development in lambda5-deficient mice. *Immunity, 5*(3), 229–239.
Pelanda, R., Schwers, S., Sonoda, E., Torres, R. M., Nemazee, D., & Rajewsky, K. (1997). Receptor editing in a transgenic mouse model: Site, efficiency, and role in B cell tolerance and antibody diversification. *Immunity*, 7(6), 765–775.
Pelanda, R., & Torres, R. M. (2012). Central B-cell tolerance: Where selection begins. *Cold Spring Harbor Perspectives in Biology*, 4(4), a007146 [Research Support, N.I.H., Extramural Research Support, Non-U.S. Gov't].
Pennell, C. A., Mercolino, T. J., Grdina, T. A., Arnold, L. W., Haughton, G., & Clarke, S. H. (1989). Biased immunoglobulin variable region gene expression by Ly-1 B cells due to clonal selection. *European Journal of Immunology, 19*(7), 1289–1295.
Perlmutter, R. M., Kearney, J. F., Chang, S. P., & Hood, L. E. (1985). Developmentally controlled expression of immunoglobulin VH genes. *Science, 227*(4694), 1597–1601.
Pewzner-Jung, Y., Friedmann, D., Sonoda, E., Jung, S., Rajewsky, K., & Eilat, D. (1998). B cell deletion, anergy, and receptor editing in "knock in" mice targeted with a germline-encoded or somatically mutated anti-DNA heavy chain. *The Journal of Immunology, 161*(9), 4634–4645.
Pogue, S. L., & Goodnow, C. C. (1994). Ig heavy chain extracellular spacer confers unique glycosylation of the Mb-1 component of the B cell antigen receptor complex. *The Journal of Immunology, 152*(8), 3925–3934.
Pritsch, O., Hudry-Clergeon, G., Buckle, M., Petillot, Y., Bouvet, J. P., Gagnon, J., et al. (1996). Can immunoglobulin C(H)1 constant region domain modulate antigen binding affinity of antibodies? *The Journal of Clinical Investigation, 98*(10), 2235–2243.
Pritsch, O., Magnac, C., Dumas, G., Bouvet, J. P., Alzari, P., & Dighiero, G. (2000). Can isotype switch modulate antigen-binding affinity and influence clonal selection? *European Journal of Immunology, 30*(12), 3387–3395.
Quach, T. D., Manjarrez-Orduno, N., Adlowitz, D. G., Silver, L., Yang, H., Wei, C., et al. (2011). Anergic responses characterize a large fraction of human autoreactive naive B cells expressing low levels of surface IgM. *The Journal of Immunology, 186*(8), 4640–4648.
Radic, M. Z., & Weigert, M. (1994). Genetic and structural evidence for antigen selection of anti-DNA antibodies. *Annual Review of Immunology, 12*, 487–520.
Ramadani, F., Bolland, D. J., Garcon, F., Emery, J. L., Vanhaesebroeck, B., Corcoran, A. E., et al. (2010). The PI3K isoforms p110alpha and p110delta are essential for pre-B cell receptor signaling and B cell development. *Science Signaling, 3*(134), ra60.
Reth, M. (1992). Antigen receptors on B lymphocytes. *Annual Review of Immunology, 10*, 97–121.
Rickert, R. C. (2013). New insights into pre-BCR and BCR signalling with relevance to B cell malignancies. *Nature Reviews Immunology, 13*(8), 578–591.
Robbins, G. R., & Knight, K. L. (2011). Mechanism for pre-B cell loss in VH-mutant rabbits. *The Journal of Immunology, 187*(9), 4714–4720.
Roes, J., & Rajewsky, K. (1993). Immunoglobulin D (IgD)-deficient mice reveal an auxiliary receptor function for IgD in antigen-mediated recruitment of B cells. *The Journal of Experimental Medicine, 177*(1), 45–55.
Rogers, K. A., Richardson, J. P., Scinicariello, F., & Attanasio, R. (2006). Molecular characterization of immunoglobulin D in mammals: Immunoglobulin heavy constant delta genes in dogs, chimpanzees and four old world monkey species. *Immunology, 118*(1), 88–100.
Rolink, A. G., Winkler, T., Melchers, F., & Andersson, J. (2000). Precursor B cell receptor-dependent B cell proliferation and differentiation does not require the bone marrow or fetal liver environment. *The Journal of Experimental Medicine, 191*(1), 23–32.

Ruminy, P., Etancelin, P., Couronne, L., Parmentier, F., Rainville, V., Mareschal, S., et al. (2011). The isotype of the BCR as a surrogate for the GCB and ABC molecular subtypes in diffuse large B-cell lymphoma. *Leukemia, 25*(4), 681–688.

Saha, N., Suetake, H., Kikuchi, K., & Suzuki, Y. (2004). Fugu immunoglobulin D: A highly unusual gene with unprecedented duplications in its constant region. *Immunogenetics, 56*(6), 438–447.

Schebesta, M., Pfeffer, P. L., & Busslinger, M. (2002). Control of pre-BCR signaling by Pax5-dependent activation of the BLNK gene. *Immunity, 17*(4), 473–485.

Schuh, W., Meister, S., Roth, E., & Jack, H. M. (2003). Cutting edge: Signaling and cell surface expression of a mu H chain in the absence of lambda 5: A paradigm revisited. *The Journal of Immunology, 171*(7), 3343–3347.

Schweighoffer, E., Vanes, L., Nys, J., Cantrell, D., McCleary, S., Smithers, N., et al. (2013). The BAFF receptor transduces survival signals by co-opting the B cell receptor signaling pathway. *Immunity, 38*(3), 475–488.

Sela-Culang, I., Alon, S., & Ofran, Y. (2012). A systematic comparison of free and bound antibodies reveals binding-related conformational changes. *The Journal of Immunology, 189*(10), 4890–4899.

Shimizu, T., Mundt, C., Licence, S., Melchers, F., & Martensson, I. L. (2002). VpreB1/VpreB2/lambda 5 triple-deficient mice show impaired B cell development but functional allelic exclusion of the IgH locus. *The Journal of Immunology, 168*(12), 6286–6293.

Simons, K., & Toomre, D. (2000). Lipid rafts and signal transduction. *Nature Reviews. Molecular Cell Biology, 1*(1), 31–39. http://dx.doi.org/10.1038/35036052.

Srinivasan, L., Sasaki, Y., Calado, D. P., Zhang, B., Paik, J. H., DePinho, R. A., et al. (2009). PI3 kinase signals BCR-dependent mature B cell survival. *Cell, 139*(3), 573–586.

Stenvik, J., & Jorgensen, T. O. (2000). Immunoglobulin D (IgD) of Atlantic cod has a unique structure. *Immunogenetics, 51*(6), 452–461.

Su, Y. W., Flemming, A., Wossning, T., Hobeika, E., Reth, M., & Jumaa, H. (2003). Identification of a pre-BCR lacking surrogate light chain. *The Journal of Experimental Medicine, 198*(11), 1699–1706.

Subrahmanyam, R., & Sen, R. (2012). Epigenetic features that regulate IgH locus recombination and expression. *Current Topics in Microbiology and Immunology, 356*, 39–63.

Suzuki, H., Terauchi, Y., Fujiwara, M., Aizawa, S., Yazaki, Y., Kadowaki, T., et al. (1999). Xid-like immunodeficiency in mice with disruption of the p85alpha subunit of phosphoinositide 3-kinase. *Science, 283*(5400), 390–392.

Takemori, T., & Rajewsky, K. (1984). Specificity, duration and mechanism of idiotype suppression induced by neonatal injection of monoclonal anti-idiotope antibodies into mice. *European Journal of Immunology, 14*(7), 656–667.

ten Boekel, E., Melchers, F., & Rolink, A. G. (1997). Changes in the V(H) gene repertoire of developing precursor B lymphocytes in mouse bone marrow mediated by the pre-B cell receptor. *Immunity, 7*(3), 357–368.

Tiegs, S. L., Russell, D. M., & Nemazee, D. (1993). Receptor editing in self-reactive bone marrow B cells. *The Journal of Experimental Medicine, 177*(4), 1009–1020.

Tokuhisa, T., Gadus, F. T., & Herzenberg, L. A. (1981). Monoclonal antibody to an IgD allotype induces a new type of allotype suppression in the mouse. *The Journal of Experimental Medicine, 154*(3), 921–934.

Tongren, J. E., Corran, P. H., Jarra, W., Langhorne, J., & Riley, E. M. (2005). Epitope-specific regulation of immunoglobulin class switching in mice immunized with malarial merozoite surface proteins. *Infection and Immunity, 73*(12), 8119–8129.

Tornberg, U. C., & Holmberg, D. (1995). B-1a, B-1b and B-2 B cells display unique VHDJH repertoires formed at different stages of ontogeny and under different selection pressures. *The EMBO Journal, 14*(8), 1680–1689.

Torres, M., Fernandez-Fuentes, N., Fiser, A., & Casadevall, A. (2007). The immunoglobulin heavy chain constant region affects kinetic and thermodynamic parameters of antibody variable region interactions with antigen. *The Journal of Biological Chemistry*, *282*(18), 13917–13927.

Treanor, B. (2012). B-cell receptor: From resting state to activate. *Immunology*, *136*(1), 21–27.

Treanor, B., Depoil, D., Gonzalez-Granja, A., Barral, P., Weber, M., Dushek, O., et al. (2010). The membrane skeleton controls diffusion dynamics and signaling through the B cell receptor. *Immunity*, *32*(2), 187–199.

Tsubata, T., & Reth, M. (1990). The products of pre-B cell-specific genes (lambda 5 and VpreB) and the immunoglobulin mu chain form a complex that is transported onto the cell surface. *The Journal of Experimental Medicine*, *172*(3), 973–976.

Tudor, D., Yu, H., Maupetit, J., Drillet, A.-S., Bouceba, T., Schwartz-Cornil, I., et al. (2012). Isotype modulates epitope specificity, affinity, and antiviral activities of anti-HIV-1 human broadly neutralizing 2F5 antibody. *Proceedings of the National Academy of Sciences*, *109*(31), 12680–12685.

Turner, M., Mee, P. J., Costello, P. S., Williams, O., Price, A. A., Duddy, L. P., et al. (1995). Perinatal lethality and blocked B-cell development in mice lacking the tyrosine kinase Syk. *Nature*, *378*(6554), 298–302.

Tze, L. E., Schram, B. R., Lam, K. P., Hogquist, K. A., Hippen, K. L., Liu, J., et al. (2005). Basal immunoglobulin signaling actively maintains developmental stage in immature B cells. *PLoS Biology*, *3*(3), e82.

Ubelhart, R., Bach, M. P., Eschbach, C., Wossning, T., Reth, M., & Jumaa, H. (2010). N-linked glycosylation selectively regulates autonomous precursor BCR function. *Nature Immunology*, *11*(8), 759–765.

Vaandrager, J. W., Schuuring, E., Kluin-Nelemans, H. C., Dyer, M. J., Raap, A. K., & Kluin, P. M. (1998). DNA fiber fluorescence in situ hybridization analysis of immunoglobulin class switching in B-cell neoplasia: Aberrant CH gene rearrangements in follicle center-cell lymphoma. *Blood*, *92*(8), 2871–2878.

Vettermann, C., Herrmann, K., Albert, C., Roth, E., Bosl, M. R., & Jack, H. M. (2008). A unique role for the lambda5 nonimmunoglobulin tail in early B lymphocyte development. *The Journal of Immunology*, *181*(5), 3232–3242.

Vettermann, C., & Schlissel, M. S. (2010). Allelic exclusion of immunoglobulin genes: Models and mechanisms. *Immunological Reviews*, *237*(1), 22–42.

Waisman, A., Kraus, M., Seagal, J., Ghosh, S., Melamed, D., Song, J., et al. (2007). IgG1 B cell receptor signaling is inhibited by CD22 and promotes the development of B cells whose survival is less dependent on Ig alpha/beta. *The Journal of Experimental Medicine*, *204*(4), 747–758 [Research Support, N.I.H., Extramural Research Support, Non-U.S. Gov't].

Wakabayashi, C., Adachi, T., Wienands, J., & Tsubata, T. (2002). A distinct signaling pathway used by the IgG-containing B cell antigen receptor. *Science*, *298*(5602), 2392–2395.

Wang, F., Ekiert, D. C., Ahmad, I., Yu, W., Zhang, Y., Bazirgan, O., et al. (2013). Reshaping antibody diversity. *Cell*, *153*(6), 1379–1393.

Wardemann, H., Yurasov, S., Schaefer, A., Young, J. W., Meffre, E., & Nussenzweig, M. C. (2003). Predominant autoantibody production by early human B cell precursors. *Science*, *301*(5638), 1374–1377.

Wasserman, R., Li, Y. S., & Hardy, R. R. (1997). Down-regulation of terminal deoxynucleotidyl transferase by Ig heavy chain in B lineage cells. *The Journal of Immunology*, *158*(3), 1133–1138.

Wedemayer, G. J., Patten, P. A., Wang, L. H., Schultz, P. G., & Stevens, R. C. (1997). Structural insights into the evolution of an antibody combining site. *Science*, *276*(5319), 1665–1669.

Wen, L., Brill-Dashoff, J., Shinton, S. A., Asano, M., Hardy, R. R., & Hayakawa, K. (2005). Evidence of marginal-zone B cell-positive selection in spleen. *Immunity*, *23*(3), 297–308.

Werner, M., Hobeika, E., & Jumaa, H. (2010). Role of PI3K in the generation and survival of B cells. *Immunological Reviews*, *237*(1), 55–71.

Wienands, J., & Reth, M. (1992). Glycosyl-phosphatidylinositol linkage as a mechanism for cell-surface expression of immunoglobulin D. *Nature*, *356*(6366), 246–248. http://dx.doi.org/10.1038/356246a0.

Wilson, M., Bengten, E., Miller, N. W., Clem, L. W., Du Pasquier, L., & Warr, G. W. (1997). A novel chimeric Ig heavy chain from a teleost fish shares similarities to IgD. *Proceedings of the National Academy of Sciences of the United States of America*, *94*(9), 4593–4597.

Yancopoulos, G. D., Desiderio, S. V., Paskind, M., Kearney, J. F., Baltimore, D., & Alt, F. W. (1984). Preferential utilization of the most JH-proximal VH gene segments in pre-B-cell lines. *Nature*, *311*(5988), 727–733.

Yang, J., & Reth, M. (2010). The dissociation activation model of B cell antigen receptor triggering. *FEBS Letters*, *584*(24), 4872–4877.

Yel, L., Minegishi, Y., Coustan-Smith, E., Buckley, R. H., Trubel, H., Pachman, L. M., et al. (1996). Mutations in the mu heavy-chain gene in patients with agammaglobulinemia. *The New England Journal of Medicine*, *335*(20), 1486–1493.

Young, R. M., & Staudt, L. M. (2013). Targeting pathological B cell receptor signalling in lymphoid malignancies. *Nature Reviews. Drug Discovery*, *12*(3), 229–243.

Zhao, Y., Kacskovics, I., Pan, Q., Liberles, D. A., Geli, J., Davis, S. K., et al. (2002). Artiodactyl IgD: The missing link. *The Journal of Immunology*, *169*(8), 4408–4416.

Zheng, N. Y., Wilson, K., Wang, X., Boston, A., Kolar, G., Jackson, S. M., et al. (2004). Human immunoglobulin selection associated with class switch and possible tolerogenic origins for C delta class-switched B cells. *The Journal of Clinical Investigation*, *113*(8), 1188–1201.

Zikherman, J., Parameswaran, R., & Weiss, A. (2012). Endogenous antigen tunes the responsiveness of naive B cells but not T cells. *Nature*, *489*(7414), 160–164.

INDEX

Note: Page numbers followed by "*f*" indicate figures.

CONTENTS OF RECENT VOLUMES

Volume 88

Volume 89

Volume 90

Volume 91

Volume 92

Volume 93

Volume 94

Volume 95

Volume 96

Volume 97

Volume 98

Volume 99

Volume 100

Volume 101

Volume 102

Volume 103

Volume 104

Volume 105

Volume 106

Volume 107

Volume 108

Volume 109

Volume 110

Volume 111

Volume 112

Volume 113

Volume 114

Volume 115

Volume 116

Volume 117

Volume 118

Volume 119

Volume 120

Volume 121

Volume 122

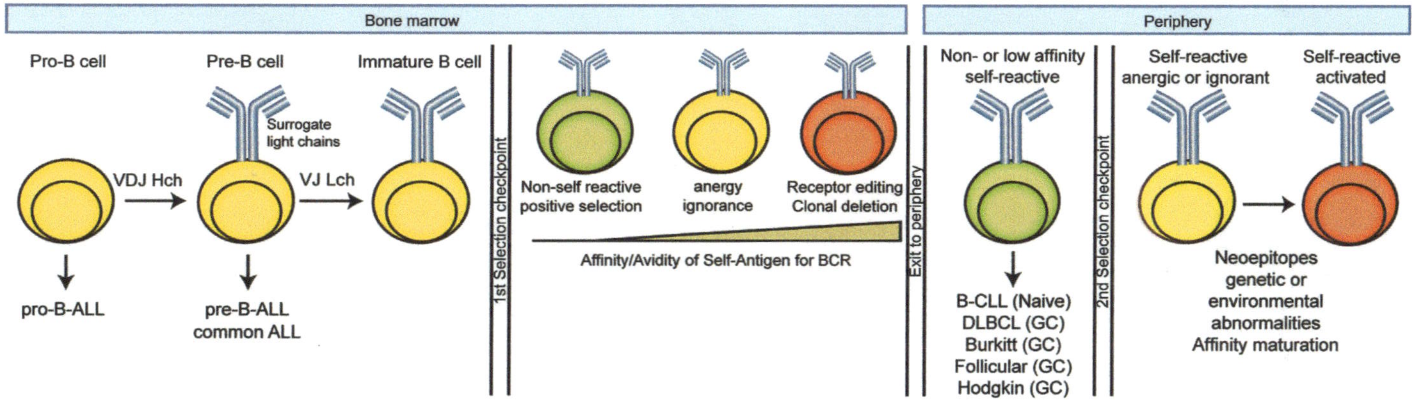

Chapter 1, Figure 1.1 (See page 7 of this volume.)

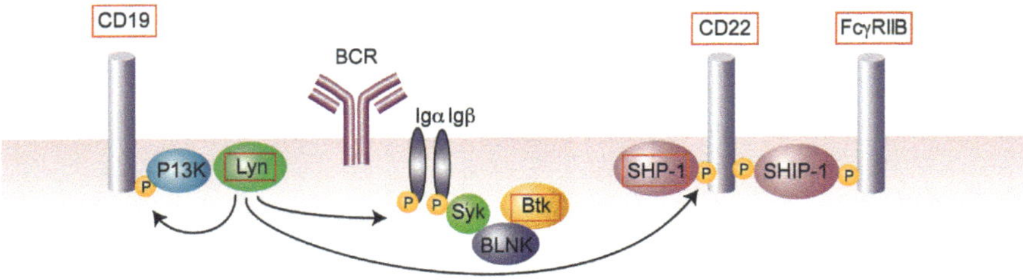

Chapter 1, Figure 1.2 (See page 18 of this volume.)

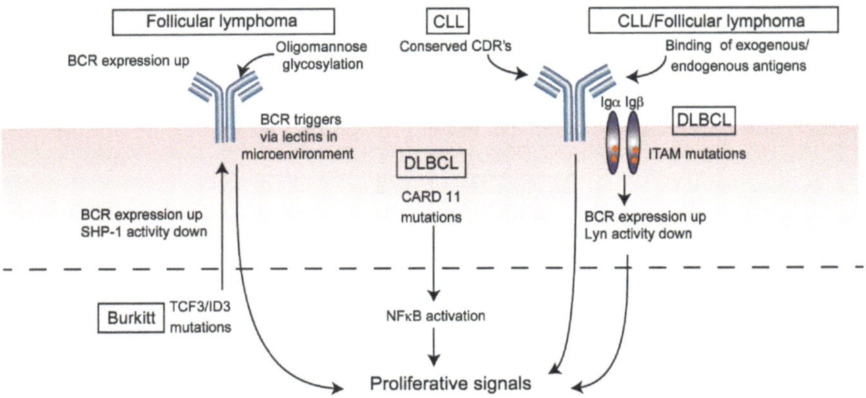

Chapter 1, Figure 1.3 (See page 21 of this volume.)

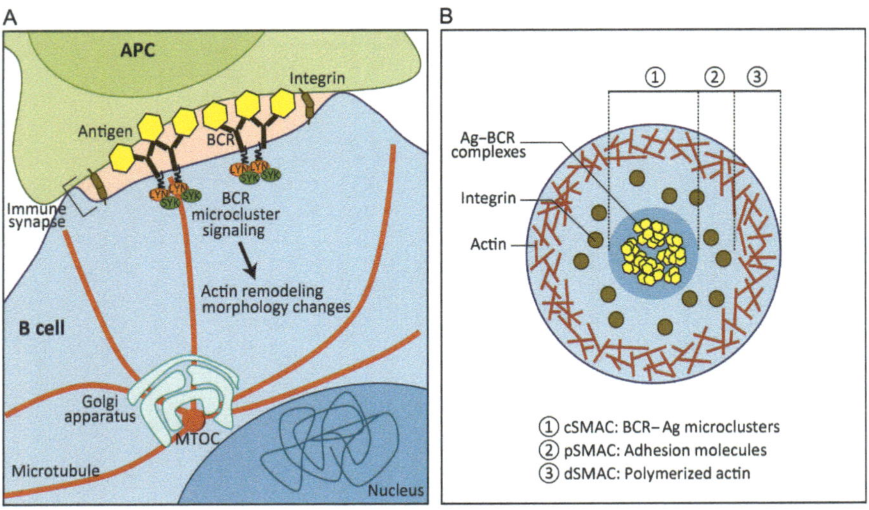

Chapter 2, Figure 2.1 (See page 54 of this volume.)

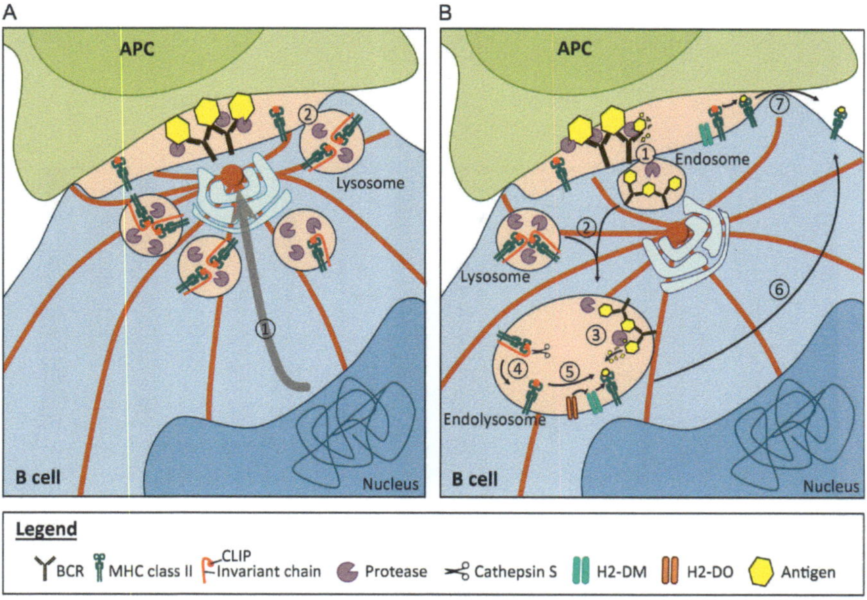

Chapter 2, Figure 2.2 (See page 56 of this volume.)

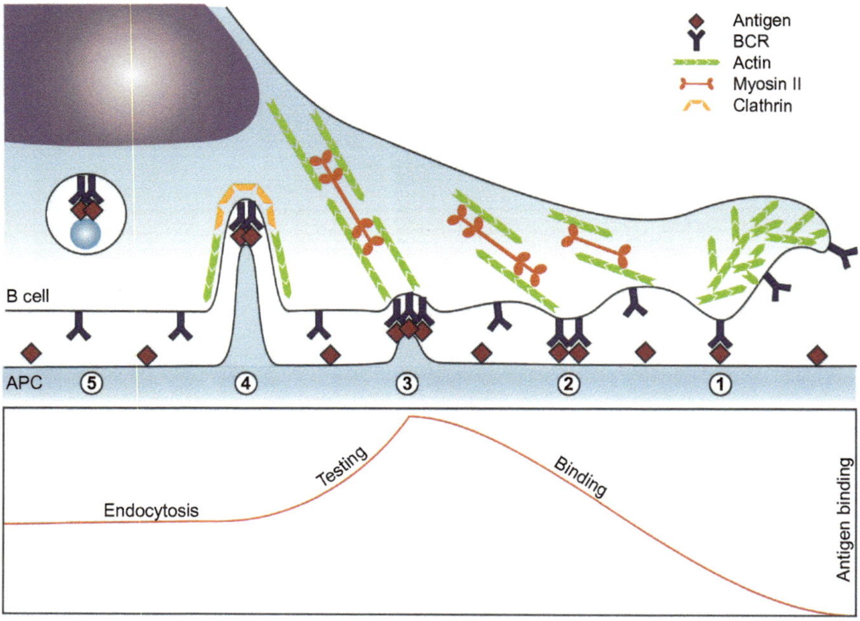

Chapter 3, Figure 3.1 (See page 73 of this volume.)

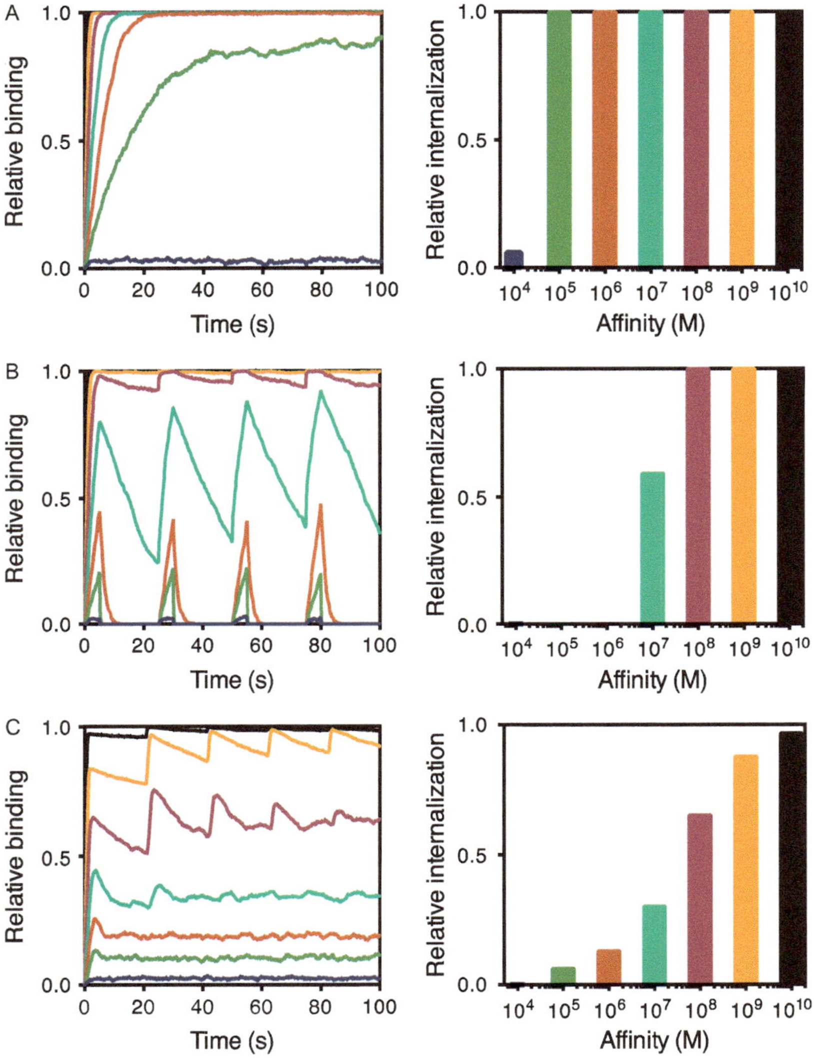

Chapter 3, Figure 3.2 (See page 87 of this volume.)

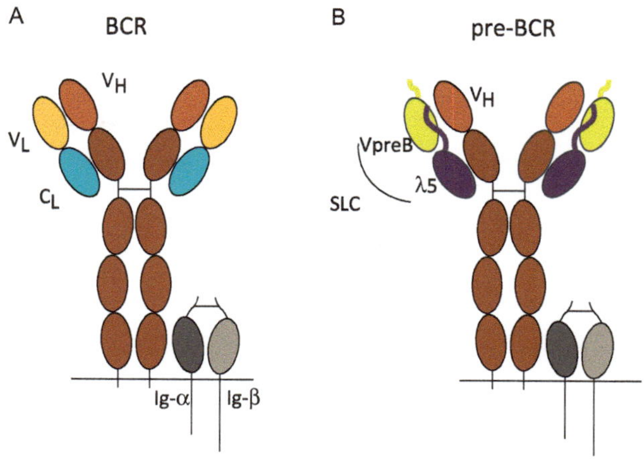

Chapter 4, Figure 4.1 (See page 102 of this volume.)

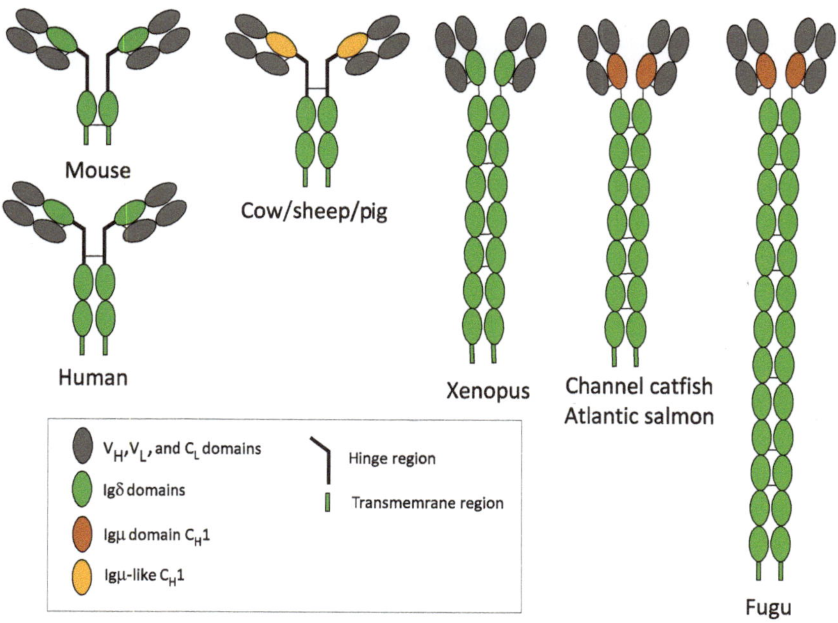

Chapter 4, Figure 4.2 (See page 119 of this volume.)

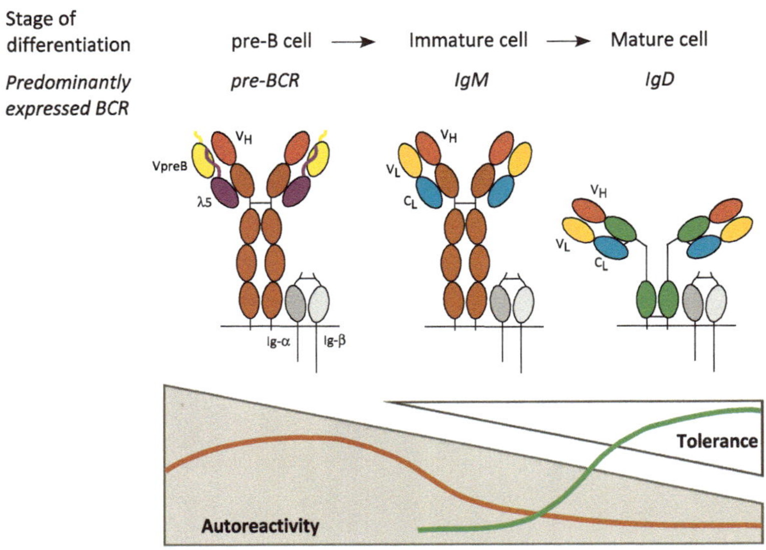

Chapter 4, Figure 4.3 (See page 122 of this volume.)

FSC
www.fsc.org
MIX
Papier | Fördert gute Waldnutzung
FSC® C083411

Kolibri 360 GmbH
Ferdinand-Jühlke-Straße 7
99095 Erfurt, Deutschland
produktsicherheit@kolibri360.de